Lecture Notes in Applied Mechanics

Volume 2

Series Editor

Prof. Dr.-Ing. Friedrich Pfeiffer

Springer

Berlin
Heidelberg
New York
Barcelona
Hong Kong
London
Milan
Paris
Singapore
Tokyo

Wave Propagation in Viscoelastic and Poroelastic Continua

A Boundary Element Approach

Martin Schanz

Springer

Dr.-Ing. MARTIN SCHANZ
Technical University Braunschweig
Institute of Applied Mechanics
Spielmannstr. 11
38106 Braunschweig
GERMANY

The present book has been accepted as Habilitation Thesis from the Department of Civil Engineering of the Technical University Carolo Wilhelmina at Braunschweig. The "venia legendi" for the scientific area of "Mechanics" was given on 9th of January 2001.

With 78 Figures and 6 Tables
ISBN 3-540-41632-3 Springer Verlag Berlin Heidelberg New York

Library of Congress Cataloging-in-Publication Data

Schanz, Martin, 1963–
Wave propagation in viscoelastic and poroelastic continua :
a boundary element approach / Martin Schanz.
p.cm. -- (Lecture notes in applied mechanics ; v. 2)
Includes bibliographical references and indexes.
ISBN 3-540-41632-3
1. Wave-motion, Theory of. 2. Viscoelasticity. 3. Boundary element methods.
I. Title. II. Series.
QA927.S218 2001
530.14--dc21 2001020508

Springer-Verlag Berlin Heidelberg New York
a member of BertelsmannSpringer Science+Business Media GmbH

http://www.springer.de

Printed in Germany

Cover design: Design & Production GmbH, Heidelberg
Typesetting: Digital data supplied by author

SPIN: 10796417 62/3020xv-5 4 3 2 1 0 – printed on acid-free paper

Preface

Wave propagation is an important topic in engineering sciences, especially, in the field of solid mechanics. A description of wave propagation phenomena is given by Graff [98]: *The effect of a sharply applied, localized disturbance in a medium soon transmits or 'spreads' to other parts of the medium.* These effects are familiar to everyone, e.g., transmission of sound in air, the spreading of ripples on a pond of water, or the transmission of radio waves. From all wave types in nature, here, attention is focused only on waves in solids. Thus, solely mechanical disturbances in contrast to electro-magnetic or acoustic disturbances are considered.

In solids, there are two types of waves – the compression wave similar to the pressure wave in fluids and, additionally, the shear wave. Due to continual reflections at boundaries and propagation of waves in bounded solids after some time a steady state is reached. Depending on the influence of the inertia terms, this state is governed by a static or dynamic equilibrium in frequency domain. However, if the rate of onset of the load is high compared to the time needed to reach this steady state, wave propagation phenomena have to be considered.

Applications of wave phenomena can be found in nearly every field of engineering. In non-destructive testing, disturbances of traveling waves are measured to identify cracks or inclusions in the material. In the field of mining, blasting induces intense stress waves to burst rocks or parts of it. Seismic waves are used to study the interior construction of the earth. In these examples, waves are artificially introduced and the wave propagation is observed. There is also an opposite application which enforce the study of wave propagation phenomena. Waves produced by earthquakes can cause tremendous destruction in buildings or other man made constructions. Therefore, knowledge is necessary how waves propagate in soil to prevent buildings or dams from destruction. Further, it is of interest to know how far and strong an impulse travels induced by a machine to prevent persons in the surrounding of such an excitation from such tremors.

This short, certainly incomplete listing shows the importance of wave propagation problems in engineering mechanics. To tackle such problems correctly will lead to an improvement of constructions and higher quality of living by protecting houses from tremors. For a small number of special problems an analytical solution is available, see, e.g., the books of Graff and Achenbach [98, 2]. But most cases, especially in two-dimensional (2-d) or three-dimensional (3-d) problems, involve numerical solutions of the governing equations based on discretization methods.

In this book, a numerical method to treat wave propagation problems in poro- and viscoelastic media is developed and tested. The method of choice is the boundary element method (BEM) since this method implicitly fulfills the Sommerfeld radiation condition. In the authors opinion, this condition is essential because many wave propagation problems include semi-infinite geometries, e.g., a half space. The crucial point in any time-dependent BEM formulation is to find time-dependent fundamental solutions. This difficulty is overcome by using the convolution quadrature method which makes it possible to establish a boundary element time-stepping procedure based on the known Laplace domain fundamental solutions for visco- or poroelastic continua. Additionally, this quadrature formula improves the stability behavior of the resulting time-stepping procedure.

After a brief literature review in chapter one, the second chapter is about the *Convolution Quadrature Method* upon which a time-dependent boundary element formulation for inelastic media is developed. Essentially, this method is a quadrature rule for the convolution integrals appearing in time-dependent integral equations. There, the fundamental solutions and the boundary data are convoluted with respect to the time variable. The advantage of the convolution quadrature method against other numerical methods to evaluate those convolution integrals is the property that the integration weights are determined using the Laplace transformed fundamental solutions. Therefore, a boundary element time-stepping procedure can be established without the knowledge of the time domain fundamental solutions. After giving the details of the convolution quadrature method some parameter studies are presented showing the sensitivity of the methodology.

In chapter three, the integral equation for a viscoelastically supported Euler-Bernoulli beam is deduced and solved with the convolution quadrature method. This is an ideal example to demonstrate the way how this quadrature rule is applied in time-dependent integral equations. Contrary to any Finite Element solutions, in this chapter an exact solution with respect to the spatial variable of an transient excited beam on a viscoelastic foundation is given. Only, the time history is approximated by the convolution quadrature method.

This solution procedure for the beam is extended to a three-dimensional boundary element formulation for an ideal elastic continuum in chapter four. After a brief description of the problem at hand, the governing time-dependent integral equation is deduced and a time-stepping boundary element formulation is established. Some parameter studies concerning the spatial and temporal discretization show the reliability of the method and give hints how to use this new method.

It is well known that beside the geometrical damping in a half space material damping has to be taken into account. The extension of Hook's law by damping terms, finally, leads to viscoelastic constitutive equations. In chapter five, these constitutive equations are introduced and the elastic-viscoelastic correspondence principle is recalled. Based on these preliminaries, the viscoelastic boundary integral equation is obtained from a generalized reciprocal work theorem. Following the discretization procedure known from the last chapter, the viscoelastic time-dependent boundary element method based on the convolution quadrature method is presented.

For this formulation, no time-dependent fundamental solutions are necessary. Numerical studies show that this formulation behaves similar to the elastodynamic formulation with respect to spatial and temporal discretization. The influence of the viscoelastic damping parameters is studied using an elastic concrete foundation slab on a viscoelastic half space.

With viscoelastic constitutive equations several materials can be described well. However, for a fluid saturated material, e.g., soil or air filled foams, a two-phase constitutive theory has to be applied. Here, Biot's poroelastic theory is used and introduced in chapter six. In the following, this theory is extended to a viscoelastic solid skeleton. After deriving fundamental solutions for poroelastic continua, the integral representation of the coupled set of differential equations for poroelastic continua is found. As in the viscoelastic case, only the Laplace domain fundamental solutions are given, and, therefore, only the convolution quadrature method makes it possible to establish a time-stepping boundary element procedure. The comparison with a semi-analytical solution of a poroelastic 1-d column shows the influence of spatial and temporal discretization. The wave propagation in a half space modeled with measured material data of soil complete this chapter.

Chapter seven deals in detail with the above mentioned semi-analytical solution of a poroelastic 1-d column. First, the analytical solution is deduced and taken to study the two compressional waves in a poroelastic material. This solution is not only needed as a reference for numerical solutions but enables us to study basic properties of a poroelastic material. In the following section, the proposed boundary element procedure is used to consider waves in a half space. The Rayleigh wave is of great interest due to its disastrous effect in earthquakes. With the established visco- and poroelastic boundary element formulations this wave is found. Its behavior is shown for viscoelastic or poroelastic media. Also, as in the case of the 1-d column, the second so-called slow compressional wave is captured.

With a concluding remark and an outlook this book is completed. The appendix with some mathematical preliminaries and the lengthy fundamental solutions listed is given for convenience.

Acknowledgments: The author has good reason to be grateful to a large number of people, and would take this opportunity of expressing his sincere gratitude to all his friends and his colleagues in the Institute of Applied Mechanics at the Technical University Braunschweig.

In particular, he gratefully acknowledges the support of Prof. Dr.rer.nat. habil. Heinz Antes throughout this work and for encouraging the application for the Habilitation degree. Thanks are also expressed to Prof. Dr.-Ing. habil. Lothar Gaul and to Prof. Dr.-Ing. habil. Wolfgang Ehlers, from the University Stuttgart, for accepting the responsibility to act as referees during the Habilitation procedure.

Finally, the author wish to thank his wife, Dr.-Ing. Karin Haese, for reading the manuscript and for many helpful suggestions.

Braunschweig, January 2001 *Martin Schanz*

Contents

1. Introduction

The most popular numerical method applied to engineering problems is the Finite Element Method (FEM). This well established method is documented in several monographs, see, e.g., [190, 21]. Applications in statics as well as in dynamics can be treated by most of the available commercial software programs. For dynamic problems, formulations in frequency and time domain exist and are well developed. Also, viscoelastic as well as quasi-static poroelastic constitutive equations are implemented in most commercial codes. Dynamic poroelastic FEM formulations were published by Zienkiewicz and Shiomi [192], Zienkiewicz et al. [191], and Diebels and Ehlers [70]. Therefore, the FEM can be used to solve wave propagation problems in viscoelastic as well as in poroelastic media, especially, in bounded domains.

However, wave propagation phenomena are often observed in semi-infinite media, e.g., earthquake motion or propagation of machine foundation excitations in the half space and their effect on neighboring buildings. Characteristically, in a semi-infinite domain only the outward propagation of waves appear and, since in infinity there is no interaction with any boundaries, no inward propagation is possible. Additionally, it is known that their effect, i.e., their amplitude, decreases with increasing distance from the point of excitation. These physical observations are mathematically formulated in the Sommerfeld radiation condition [171]. A suitable numerical method for calculating wave propagation in semi-infinite media has to ensure that this condition is not violated.

In the FEM, special techniques have to be applied to fulfill this condition. There are two main ideas to solve this problem: first, to use the so-called infinite elements [27] or, second, to use the so-called Dirichlet-to-Neumann boundary condition [106, 97]. A recently published book gives an overview on such methods [95]. But, using the boundary element method (BEM) the Sommerfeld radiation condition is implicitly fulfilled. This discretization method is based, as the FEM, on a weighted residual formulation of the governing differential equation but, contrary to the FEM, as weighting functions fundamental solutions are used instead of the variation of the ansatz functions. This advantage is one of the main reasons to use the BEM. Therefore, in the following, the review is restricted to the BEM.

BEM in elasticity

The BEM considered as an integral equation method has a long history that can go back to 1903 when Fredholm [87] published his rigorous work on integral equations encountered in potential theory. The classical works of Kellogg [112] in 1929 on potential theory and, especially, of Muskhelishvili [131, 132] in 1953 and Kupradze [116] in 1965 on elastostatics represent applications of integral equation techniques. The term BEM, which first appeared in the literature in 1977 in the works of Banerjee and Butterfield [19] and Brebbia and Domínguez [40], indicates the surface discretization character of the method. Beside some indirect BE formulations, the direct BEM was first introduced in an explicit and general way by Jawson [107] in 1963 in connection with the potential theory and by Rizzo [151] in 1967 and Cruse [60] in 1969 in connection with elastostatics. A more extensive historical review and applications of boundary integral methods may be found in the article by Beskos [25]. An introduction to BEM is given in several standard text books, e.g., by Banerjee and Butterfield [20] or by Brebbia et al. [41].

To treat wave propagation problems dynamic formulations are necessary, i.e., inertia terms have to be taken into account. The first boundary integral formulation for elastodynamics was published by Cruse and Rizzo [61, 59]. This formulation performs in Laplace domain with a subsequent inverse transformation to time domain to achieve results for transient behavior. The corresponding formulation in Fourier domain, i.e., frequency domain, was presented by Domínguez [72, 73]. The first boundary element formulation directly in the time domain was developed by Mansur for the scalar wave equation and for elastodynamics with zero initial conditions [126, 127]. The extension of this formulation to non-zero initial conditions was presented by Antes [4]. Detailed information about this procedure may be found in the book of Domínguez [76]. A comparative study of these possibilities to treat elastodynamic problems with BEM is given by Manolis [122]. A completely different approach to handle dynamic problems utilizing static fundamental solutions is the so-called dual reciprocity BEM. This method was introduced by Nardini and Brebbia [134] and details may be found in the monograph of Partridge et al. [141]. A very detailed review on elastodynamic boundary element formulations and a list of applications, e.g., soil-structure-interaction problems [18], the dynamic analysis of 3-d foundations [110], or contact problems [11], can be found in two articles of Beskos [24, 23].

The above listed methodologies to treat elastodynamic problems with the BEM show mainly the two ways: direct in time domain or via an inverse transformation in Laplace domain. Mostly, the latter is used, e.g., [3]. Since all numerical inversion formulas depend on a proper choice of their parameters [133, 53], a direct evaluation in time domain seems to be preferable. Also, it is more natural to work in the real time domain and observe the phenomenon as it evolves. But, as all time-stepping procedures, such a formulation requires an adequate choice of the time step size. An improper chosen time step size leads to instabilities or numerical damping. Four procedures to improve the stability of the classical dynamic time-stepping BE formulation can be quoted: the first employs modified numerical time marching procedures, e.g., Antes and Jäger [10], Yu et al. [187] for acoustics, Schanz et al. [166], Peirce and Siebrits [142, 33], Yu et al. [186], Schanz and Antes [161] for elastodynamics; the second employs a modified fundamental solution, e.g., Rizos and

Karabalis [150], Coda and Venturini [56] for elastodynamics; the third employs an additional integral equation for velocities [128]; and the last uses weighting methods, e.g., Yu et al. [188] for elastodynamics and Yu et al. [189] for acoustics.

However, none, with one exception, of the above direct time domain formulation could be extended to inelastic, i.e., visco- or poroelastic, material laws since no closed form fundamental solutions for such materials exist. The exception is the formulation presented originally by Schanz and Antes in [161] and discussed in detail here in Chap. 4. This formulation is based on the convolution quadrature method proposed by Lubich [118, 119]. It utilizes the Laplace domain fundamental solution and results not only in a more stable time stepping procedure but it makes it possible to take damping effects in case of visco- or poroelasticity into account.

Viscoelasticity

First viscoelastic theories are found in 1874 by Boltzmann [34] and Meyer [130] simultaneously for isotropic viscoelastic media. Based on these theories, nowadays, two ways describing viscoelastic stress-strain relations are used: i) an integral equation of hereditary type with relaxation functions or creep functions as kernels following Boltzmann, or ii) a differential equation following Meyer. From the variety of monographs on viscoelasticity, Christensen [55] presents the linear theory and includes the solution of advanced problems of research interest. Gurtin and Sternberg [102] present a postulational approach to the linear theory, emphasizing the proofs of theorems. The work of Flügge [86] is an early introduction of the linear theory, whereas Lakes [117] combines theory and experiments in his recently published book.

Improved curve fitting of measured material properties by constitutive equations with fewer parameters is achieved with the concept of fractional differ-integration (on fractional calculus see, e.g., [137, 145]). First works on viscoelastic constitutive equations with fractional derivatives have been presented by Caputo [43] and Torvik and Bagley [176, 16]. An application of these constitutive equations in case of a single degree of freedom system is given by Gaul et al. [88]. Also, an one-dimensional (1-d) problem has been treated by Drozdov [77] using measured and fitted material data. A very extensive review about fractional calculus applied to dynamic problems has recently been given by Rossikhin and Shitikova [152].

Viscoelastic boundary element formulations are mostly published for the quasi-static case (see, e.g., [175, 170, 44]), or in dynamics using a frequency or Laplace domain representation of the governing integral equation. These formulations are developed by applying the elastic-viscoelastic correspondence principle to the elastodynamic boundary element formulation, e.g., a frequency domain formulation by Kobayashi and Kawakami [115] or a Laplace domain formulation by Manolis and Beskos [123]. In these formulations, the complex moduli concept is used which can be extended to fractional operator viscoelasticity allowing not only integer powers of the frequency, see Gaul et al. [89].

Calculation of transient response, however, requires the inverse transformation. As mentioned above, numerical inversion formulas depend on a proper choice of their parameters [133], and transient boundary conditions, e.g., contact, can not be taken into account. But, formulations directly in time domain require the knowledge of viscoelastic fundamental solutions which are not yet known for the general viscoelastodynamic case. Only for a simple Maxwell model, a solution has been obtained analytically by Gaul and Schanz [90] and has been implemented in a boundary element formulation [165]. Based on the frequency domain fundamental solution with subsequent inverse transformation, a 1-d solution has been proposed by Wolf and Dabre [185]. In the 3-d case, Gaul and Schanz [92] developed a formulation for a generalized (with fractional derivatives) three-parameter model using the Laplace transformed fundamental solution which is inverted within each time step. Recently, Schanz and Antes [162] published a viscoelastic formulation based on the convolution quadrature method. This formulation takes advantage of the quadrature formula which integration weights are determined by the existing Laplace transformed fundamental solution. A complete and detailed description of this procedure is given Chap. 5. A comparison of the above mentioned two time domain formulations and the Laplace transformed formulation with a subsequent inverse transformation was presented by Gaul and Schanz [94].

Poroelasticity

For a wide range of fluid infiltrated materials, such as water saturated soils, oil impregnated rocks, or air filled foams, the elastic theory and also a viscoelastic description of the material behavior is a crude approximation for investigating wave propagation in such media. Due to their porosity, a different theory is necessary.

A historical view on this subject identifies two theories which have been developed and are used nowadays. For more details, the reader is directed to the work of de Boer, see [63, 64] or the recently published book [62]. The first works on porous media are attributed to Fillunger in 1913 [85]. In this paper and in subsequent ones, Fillunger was concerned with the question of buoyancy of barrages. Another, more intuitively theory, has been developed by von Terzaghi in 1923 [181]. These two basic works form the basis of two theories used up to day.

Based on the work of von Terzaghi, a theory of porous materials containing a viscous fluid was presented by Biot [28]. This has generally been attributed as the starting point of the theory of Poroelasticity. In the following years, Biot extended his theory to anisotropic cases [29] and also to poroviscoelasticity [30]. The dynamic extension was published in two papers, one for low frequency range [31] and the other for high frequency range [32]. Among the significant findings was the identification of three waves for a 3-d continuum, two compressional waves and one shear wave. This extra compressional wave, known as the slow wave, has been experimentally confirmed [144]. In Biot's theory a fully saturated material is assumed. The extension to a nearly saturated poroelastic solid was presented by Vardoulakis and Beskos [179].

Based on the work of Fillunger, a different approach, the Theory of Porous Media, has been developed. This theory is based on the axioms of continuum theories of mixtures [39, 178] extended by the concept of volume fractions by

Ehlers [80, 82, 81], thus proceeding from the assumption of immiscible and superimposed continua with internal interactions. It has been demonstrated that under small deformations, and some other restrictions, this and Biot's theory lead to the same governing equations [83]. Although Biot's theory is more based on physical intuition, it has the widest acceptance in geophysics and geomechanics.

Independent of which formulation is chosen, the governing equations consist of a system of coupled partial differential equations. To find a close form exact solution for the general material case, even in a simple 1-d geometry, has so far not been successful. Some analytical solutions for special 1-d problems have been found. For example, Grag et al. [100] examined the response of an infinitely long fluid saturated soil column subjected to a Heaviside step function velocity boundary condition at one end. A solution in frequency domain of a finite 1-d column loaded at the top by total stress and pore pressure was presented in Cheng et al. [48] for comparison with a boundary element solution. The corresponding time-dependent solution is given by Schanz and Cheng [164] (see Sect. 7.1). All of the above mentioned solutions assume a fully saturated material using Biot's theory. For an even more general material case of a partially saturated dual-porosity medium, a 1-d solution in Laplace domain is available from Vgenopoulou and Beskos [180]. The Theory of Porous Media was used to solve analytically the problem of an infinitely long column with incompressible constituents [65].

Apart from these one-dimensional solutions, in general, a numerical method has to be applied. A two-dimensional quasi-static BE formulation has been developed by Cheng and Ligget for consolidation problems [51] and for fracture [50]. Later, a three-dimensional quasi-static formulation was published by Badmus et al. [15]. A complete overview on the different available quasi-static formulations can be found in [49]. Those and also the following BE formulations are based on Biot's theory. For the Theory of Porous Media no boundary element formulation exists since no fundamental solutions have been found.

In case of dynamic BE formulations, the situation is similar to viscoelasticity. Since no closed form time-dependent fundamental solution is available, first poroelastodynamic BE formulations based on Biot's theory have been published in Laplace domain by Manolis and Beskos [124] expressed in terms of solid and fluid displacements. However, it can be shown that only the solid displacements and one additional variable, the fluid pressure, are independent [35]. Based on these four (three) unknowns in 3-d (2-d) formulations in frequency domain have been published by Cheng et al. [48] and Domínguez [75]. In these formulations, the transient response of a poroelastic continuum has be determined with an inverse transformation. As discussed above, to work in the real time domain is preferable also in case of poroelasticity. Such a time domain formulation was developed by Wiebe and Antes [184], but with the restriction of vanishing damping between the solid skeleton and the fluid. Another time dependent formulation was proposed by Chen and Dargush based on analytical inverse transformation of the Laplace domain fundamental solutions [47], but, as the author admits, this formulation is highly CPU-time demanding. Recently, the author developed a poroelastic time stepping

BE procedure similar to the viscoelastic case based on the convolution quadrature method [160, 159, 158] which is detailed discussed in Chap. 6.

2. Convolution quadrature method

In many engineering applications the convolution integral

$$y(t) = f(t) * g(t) = \int_0^t f(t-\tau)\, g(\tau)\, \mathrm{d}\tau \tag{2.1}$$

plays a crucial role. A standard situation is to find the response of a system corresponding to an arbitrary time-dependent load if the impulse response function is known. The convolution between the impulse response function and the loading is the solution. Other examples of convolution integrals are integral equations in time domain for elastodynamics or the hereditary integral formulation of viscoelastic constitutive equations.

Mostly, the convolution (2.1) has to be performed numerically, because either the functions are to complicated or results from other numerical methods and, therefore, not known in closed form. A special quadrature rule for convolution integrals the *Convolution Quadrature Method* was developed by Lubich [118, 119]. This method numerically approximates the convolution integral (2.1) by a quadrature rule whose weights are determined by the Laplace transformed function $\hat{f}$ and a linear multistep method. This approximation has been applied, e.g., to the Helmholtz equation [120], to the integral equation for the heat equation [121], and to boundary element formulations for elastodynamic [156], viscoelastic [157], and poroelastic continua [160].

2.1 Basic theory of the convolution quadrature method

Substituting $f(t)$ by the inverse Laplace transformation of $\hat{f}(s)$ (see appendix A.3) in the convolution integral (2.1) and exchanging the integrals leads to

$$\int_0^t f(t-\tau)\, g(\tau)\, \mathrm{d}\tau = \frac{1}{2\pi \mathrm{i}} \lim_{R\to\infty} \int_{c-\mathrm{i}R}^{c+\mathrm{i}R} \hat{f}(s) \underbrace{\int_0^t \mathrm{e}^{s(t-\tau)} g(\tau)\, \mathrm{d}\tau}_{x(t,s)}\, \mathrm{d}s \qquad t > 0 \tag{2.2}$$

with a real constant c. The inner integral, abbreviated with $x(t,s)$, is a solution of the differential equation of first order

$$\frac{\mathrm{d}}{\mathrm{d}t}x(t,s) = sx(t,s) + g(t) \quad \text{with} \quad x(0,s) = 0\,. \tag{2.3}$$

After discretizing the time t in N equal time steps Δt the convolution at discrete times $t = n\Delta t$

$$y(n\Delta t) = \frac{1}{2\pi \mathrm{i}} \lim_{R\to\infty} \int_{c-\mathrm{i}R}^{c+\mathrm{i}R} \hat{f}(s)\, x(n\Delta t, s)\, \mathrm{d}s\,, \quad \text{with} \quad n = 0,1,\ldots,N \tag{2.4}$$

is given. The solution of (2.3) can be approximated by a linear multistep method

$$\begin{aligned} &\alpha_k x_{n+k} + \alpha_{k-1} x_{n+k-1} + \ldots + \alpha_0 x_n \\ &\qquad = \Delta t \left[\beta_k \left(s x_{n+k} + g\left((n+k)\Delta t\right)\right) + \ldots + \beta_0 \left(s x_n + g(n\Delta t)\right)\right] \end{aligned} \tag{2.5}$$

with $x(n\Delta t, s) \approx x_n$ denoting the approximated solution of $x(t,s)$ at the discrete time $n\Delta t$. This representation of a multistep method is not suitable to extract the discrete values x_n to be inserted in equation (2.4). Therefore, equation (2.5) is multiplied with z^n $(z \in \mathbb{C})$ and summed up over n from 0 to ∞

$$\begin{aligned} &\sum_{n=0}^{\infty} \alpha_k x_{n+k} z^n + \ldots + \sum_{n=0}^{\infty} \alpha_0 x_n z^n \\ &\quad = \Delta t \sum_{n=0}^{\infty} \left[\beta_k \left(s x_{n+k} + g\left((n+k)\Delta t\right)\right) z^n + \ldots + \beta_0 \left(s x_n + g(n\Delta t)\right) z^n\right]. \end{aligned} \tag{2.6}$$

Assuming vanishing starting values $x_0 = \ldots = x_{k-1} = 0$ the sum $\sum_{n=0}^{\infty} x_{n+k} z^n$ is modified

$$\sum_{n=0}^{\infty} x_{n+k} z^n = z^{-k} \sum_{n=0}^{\infty} x_{n+k} z^{n+k} = z^{-k} \sum_{m=k}^{\infty} x_m z^m = z^{-k} \sum_{n=0}^{\infty} x_n z^n\,. \tag{2.7}$$

Under the same assumption for $g(t<0) = 0$ also $\sum_{n=0}^{\infty} g\left((n+k)\Delta t\right) z^n$ is modified following equation (2.7). Next, with the property (2.7) the sums $\sum_{n=0}^{\infty} x_n z^n$ and $\sum_{n=0}^{\infty} g(n\Delta t) z^n$ can be factored out

$$\begin{aligned} &\left[\alpha_0 + \alpha_1 z^{-1} + \ldots + \alpha_k z^{-k}\right] \sum_{n=0}^{\infty} x_n z^n \\ &\qquad = \Delta t \left[\beta_0 + \ldots + \beta_k z^{-k}\right] \left[s \sum_{n=0}^{\infty} x_n z^n + \sum_{n=0}^{\infty} g(n\Delta t) z^n\right]. \end{aligned} \tag{2.8}$$

The quotient

$$\gamma(z) = \frac{\alpha_0 + \ldots + \alpha_k z^{-k}}{\beta_0 + \ldots + \beta_k z^{-k}} = \frac{\alpha_0 z^k + \ldots + \alpha_k}{\beta_0 z^k + \ldots + \beta_k} \tag{2.9}$$

characterizes the underlying multistep method (2.5) and is, therefore, in the following called characteristic function. Using the characteristic function $\gamma(z)$ and rearranging (2.8) the function $x(t,s)$ is represented in a formal power series

$$\sum_{n=0}^{\infty} x_n z^n = \frac{1}{\frac{\gamma(z)}{\Delta t} - s} \sum_{n=0}^{\infty} g(n\Delta t) z^n . \tag{2.10}$$

Before this representation of $x(n\Delta t, s)$ can be inserted in (2.4) there, also, the multiplication with z^n and summation over n has to be done. Finally, in the convolution integral

$$\sum_{n=0}^{\infty} y(n\Delta t) z^n = \frac{1}{2\pi \mathrm{i}} \lim_{R\to\infty} \int_{c-iR}^{c+iR} \hat{f}(s) \frac{1}{\frac{\gamma(z)}{\Delta t} - s} \mathrm{d}s \sum_{n=0}^{\infty} g(n\Delta t) z^n \tag{2.11}$$

remains only the complex integral along the parallel line to the imaginary axis located at c. This integration path is changed to a closed contour shown in Fig. 2.1. To

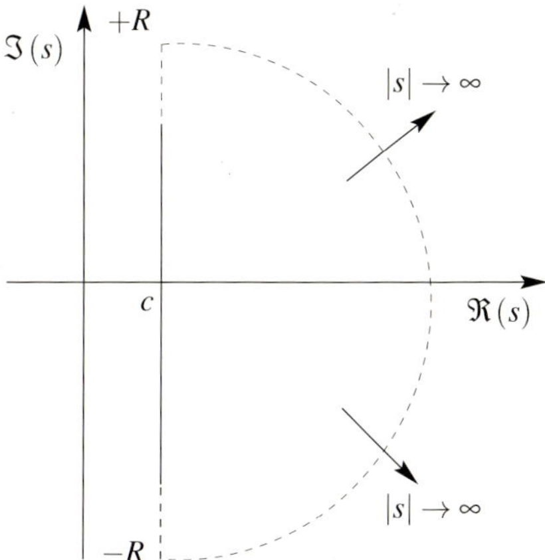

Fig. 2.1. Closed contour of integration path

do so, the function $\hat{f}$ has to fulfill the requirements

$$|\hat{f}(s)| \to 0 \quad \text{for} \quad \Re(s) \geq c \quad \text{and} \quad |s| \to \infty , \tag{2.12}$$

i.e., the function $\hat{f}$ is bounded at infinity and, therefore, the integral over the half circle vanishes for $R \to \infty$. Now, the integration can be performed with the residue theorem. Due to the definition of the inverse Laplace transform all singularities of the function $\hat{f}$ must be in the left half plane of the line located at c. Then the only remaining singularity of the integrand in (2.11) inside the contour is $\gamma(z)/\Delta t = s$, which is of order one. Following the residue theorem [1], the complex integral is determined by

$$\begin{aligned} \sum_{n=0}^{\infty} y(n\Delta t) z^n &= \frac{1}{2\pi \mathrm{i}} \lim_{R\to\infty} \int_{c-\mathrm{i}R}^{c+\mathrm{i}R} \hat{f}(s) \frac{1}{\frac{\gamma(z)}{\Delta t} - s} \mathrm{d}s \sum_{n=0}^{\infty} g(n\Delta t) z^n \\ &= \hat{f}\left(\frac{\gamma(z)}{\Delta t}\right) \sum_{n=0}^{\infty} g(n\Delta t) z^n . \end{aligned} \tag{2.13}$$

Now, remember our goal was to find an expression for $y(n\Delta t)$ not for the sum over it as in (2.13). For this, it is necessary to represent the right hand side of (2.13) in a series with coefficients independent from z.

Therefore, in the next step $\hat{f}(\gamma(z)/\Delta t)$ is developed in a power series

$$\hat{f}\left(\frac{\gamma(z)}{\Delta t}\right) = \sum_{n=0}^{\infty} \omega_n(\Delta t)\, z^n \,. \tag{2.14}$$

The coefficients $\omega_n(\Delta t)$ of the power series (2.14) are determined either

1. by an comparison of the coefficients, if $\hat{f}\left(\frac{\gamma(z)}{\Delta t}\right)$ can be rearranged as a series.
2. or for arbitrary functions $\hat{f}\left(\frac{\gamma(z)}{\Delta t}\right)$ by Cauchy's integral formula

$$\omega_n(\Delta t) = \frac{1}{2\pi \mathrm{i}} \int\limits_{|z|=\mathscr{R}} \hat{f}\left(\frac{\gamma(z)}{\Delta t}\right) z^{-n-1} \mathrm{d}z \tag{2.15}$$

with $\mathscr{R}$ being the radius of a circle in the domain of analyticity of $\hat{f}(\gamma(z)/\Delta t)$.

With the power series (2.14) a double sum appears which is simplified by the Cauchy product of two series [42]

$$\sum_{n=0}^{\infty} \omega_n(\Delta t)\, z^n \sum_{n=0}^{\infty} g(n\Delta t)\, z^n = \sum_{n=0}^{\infty}\sum_{k=0}^{n} \omega_{n-k}(\Delta t)\, g(k\Delta t)\, z^n . \tag{2.16}$$

Inserting this double sum (2.16) in (2.13) gives

$$\sum_{n=0}^{\infty} y(n\Delta t)\, z^n = \hat{f}\left(\frac{\gamma(z)}{\Delta t}\right) \sum_{n=0}^{\infty} g(n\Delta t)\, z^n = \sum_{n=0}^{\infty}\sum_{k=0}^{n} \omega_{n-k}(\Delta t)\, g(k\Delta t)\, z^n \,. \tag{2.17}$$

A comparison of coefficients results in the final *quadrature rule* for *convolution integrals*

$$y(n\Delta t) = \sum_{k=0}^{n} \omega_{n-k}(\Delta t)\, g(k\Delta t)\,, \quad n = 0, 1, \ldots, N \,. \tag{2.18}$$

Equation (2.18) is an approximation of the convolution integral (2.1) which is only based on the Laplace transformed function $\hat{f}(s)$ and the discrete values of the other function in (2.1) $g(k\Delta t)$. This property can especially be exploited to solve integral equations in time domain, e.g., [162]. Moreover, for short time periods formula (2.18) is even superior to other inverse Laplace transform methods [164].

The only approximation used to deduce the quadrature rule (2.18) is the linear multistep method (2.5) whereas all other calculations are exact. In the original work of Lubich [118] it is shown that the multistep method must be *consistent* and *strongly zero stable*. Further, the multistep method must be A-stable and $\hat{f}(s)$ be bounded in the right half plane from the line $(c - \mathrm{i}\infty, c + \mathrm{i}\infty)$, i.e.,

$$|\hat{f}(s)| \leq K \cdot |s|^{-\mu} \quad \text{for} \quad K < \infty\,, \; \mu > 0\,. \tag{2.19}$$

However, if the function $\hat{f}(s)$ is analytical and bounded in the area $|\arg(s-c)| < \pi - \phi$ with $\phi < \frac{\pi}{2}$, the stability criterium can be weakened to $A(\alpha)$-stable methods. Expressed in terms of the characteristic function $\gamma(z)$ this reads

- $\gamma(z)$ has neither zeros nor poles on the closed unit disk $(|z| \leq 1)$, (2.20a) with the exception of a single zero at $|z| = 1$,
- $|\arg \gamma(z)| \leq \pi - \alpha$, with $|z| \leq 1$, for $\alpha > \phi$, (2.20b)
- $\Delta t^{-1} \gamma\left(e^{-\Delta t}\right) = 1 + \mathscr{O}\left(\Delta t^p\right)$, with $\Delta t \to 0$ for $p \geq 1$. (2.20c)

Well known examples of multistep methods which fulfill these requirements are the *Backward Differential Formulas* (BDF) of order $p \leq 6$, e.g., BDF 2 of order two which is even A-stable ($\alpha = 90^o$), $\gamma(z) = 3/2 - 2z + z^2/2$.

As mentioned above, the only approximation introduced until now is the linear multistep method for the approximation of $x(t,s)$. If, however, the function $\hat{f}(\gamma(z)/\Delta t)$ can not analytically be developed in a power series, the coefficients of the power series must be calculated by Cauchy's integral formula (2.15). In general, the computation of the integral in (2.15) for the determination of the integration weights ω_n must be performed numerically. To do this, a polar coordinate transformation $z = \mathscr{R}\mathrm{e}^{\mathrm{i}\varphi}$ is introduced in (2.15)

$$\omega_n(\Delta t) = \frac{\mathscr{R}^{-n}}{2\pi} \int_0^{2\pi} \hat{f}\left(\frac{\gamma\left(\mathscr{R}\mathrm{e}^{\mathrm{i}\varphi}\right)}{\Delta t}\right) \mathrm{e}^{-\mathrm{i}n\varphi} \mathrm{d}\varphi \tag{2.21}$$

changing the complex in a real valued integral. Then this integral can be approximated by the trapezoidal rule

$$\omega_n(\Delta t) = \frac{\mathscr{R}^{-n}}{L} \sum_{\ell=0}^{L-1} \hat{f}\left(\frac{\gamma\left(\mathscr{R}\mathrm{e}^{\mathrm{i}\ell\frac{2\pi}{L}}\right)}{\Delta t}\right) \mathrm{e}^{-\mathrm{i}n\ell\frac{2\pi}{L}}\,, \tag{2.22}$$

with L equal steps $2\pi/L$.

The mathematical proofs of convergence and stability can be found in the papers [118, 121, 120] for different kind of applications. Assuming that $\hat{f}(s)$ in equation (2.22) is computed with an error bounded by ε, the choice $L = N$ and $\mathscr{R}^N = \sqrt{\varepsilon}$ yields an error in ω_n of size $\mathscr{O}\left(\sqrt{\varepsilon}\right)$ [119]. Additionally, the choice $L = N$ has the advantage that the N weights ω_n can be calculated very fast with the technique of the Fast Fourier Transformation (FFT) using only $\mathscr{O}(N \log N)$ operations instead of $\mathscr{O}\left(N^2\right)$.

2.2 Numerical tests

As mentioned in the last section, the integration weights in the convolution quadrature method (2.18) are determined using formula (2.22). In this formula, two parameters L and ε are free and have to be adjusted to achieve proper results. Here, some

parameter studies will reveal their influence on the accuracy of the solution. For this, the numerical results of two convolution integrals achieved by the convolution quadrature method (2.18) will be compared with known analytical solutions. For the following test, especially, functions appearing in wave propagation problems are used as test functions. The jumps at wave fronts are usually expressed by the Dirac distribution $\delta(t)$ and the Heaviside function $H(t)$ and the transient excitation are usual also Heaviside functions. Therefore, the two convolution integrals
Integral 1:

$$\left.\begin{array}{l} f_1(t) = \delta(t-a) \\ g_1(t) = (H(t) - H(t-b)) \end{array}\right\} \Rightarrow f_1(t) * g_1(t) = H(t-a) - H(t-(a+b)) \tag{2.23}$$

Integral 2:

$$\left.\begin{array}{l} f_2(t) = tH(t-a) \\ g_2(t) = H(t) \end{array}\right\} \Rightarrow f_2(t) * g_2(t) = \frac{1}{2}\left(t^2 - a^2\right) H(t-a) , \tag{2.24}$$

are used for the test.

The accuracy of the proposed quadrature rule (2.18) is influenced by the underlying multistep method characterized by $\gamma(z)$ and by the accuracy of the integration weights ω_n. In Sect. 2.1, two possibilities to determine the integration weights are given. Either representing $\hat{f}(\gamma(z)/\Delta t)$ analytically in a series or taking formula (2.22) for the determination of the coefficients of the power series. The second possibility will be normally used because for most functions $\hat{f}$ an analytical representation as a series is not possible. But, using formula (2.22) introduces another approximation. Lubich suggested in [119] the choice $\mathscr{R}^N = \sqrt{\varepsilon}$ and $L = N$ to get an error of order $\mathscr{O}(\sqrt{\varepsilon})$ and in [121] $L = 2N$. Here, results from different values for L are compared with the results using the analytical series expansion which is possible for the test functions f_1 in (2.23) and f_2 in (2.24).

2.2.1 Series expansion of the test functions f_1 and f_2

The Laplace transform of f_1 is given (see, e.g., [71])

$$\delta(t-a) \circ\!\!-\!\!\bullet\, \mathrm{e}^{-as} \tag{2.25}$$

where the Laplace variable s must be exchanged by $\gamma(z)/\Delta t$ for the series expansion. In the next step, a linear multistep method must be chosen. Here, a BDF 2, $\gamma(z) = 3/2 - 2z + z^2/2$, will be used. The influence of the parameters which will be studied in the following would be the same for a BDF 1, $\gamma(z) = 1 - z$.

The series expansion for the BDF 2 is not obvious, because in

$$\begin{aligned} \hat{f}\left(\frac{\gamma(z)}{\Delta t}\right) &= \mathrm{e}^{-a\frac{\frac{3}{2}-2z+\frac{1}{2}z^2}{\Delta t}} = \mathrm{e}^{-a\frac{3}{2\Delta t}}\,\mathrm{e}^{\frac{a}{\Delta t}\left(2z-\frac{1}{2}z^2\right)} \\ &= \mathrm{e}^{-a\frac{3}{2\Delta t}} \sum_{n=0}^{\infty} \left(\frac{a}{\Delta t}\right)^n \frac{z^n}{n!}\left(2-\frac{1}{2}z\right)^n \end{aligned} \tag{2.26}$$

z^n can not be extracted directly. However, in general, taking a power series to the power of n can be calculated by [153]

$$\left(\sum_{k=0}^{\infty} a_k z^k\right)^n = \sum_{k=0}^{\infty} c_k z^k\,, \quad c_0 = a_0^n,\; c_k = \frac{1}{k a_0}\sum_{i=1}^{k}(in - k + i)\,a_i c_{k-i}\,. \tag{2.27}$$

The last expression between the brackets in equation (2.26) is a power series with the coefficients $a_0 = 2, a_1 = -1/2$ and $a_i = 0,\ i = 2, 3, \ldots, \infty$. Consequently, formula (2.27) can be applied and leads to

$$\mathrm{e}^{-a\frac{3}{2\Delta t}} \sum_{n=0}^{\infty} \left(\frac{a}{\Delta t}\right)^n \frac{z^n}{n!}\left(2 - \frac{1}{2}z\right)^n = \mathrm{e}^{-a\frac{3}{2\Delta t}} \sum_{n=0}^{\infty} \left(\frac{a}{\Delta t}\right)^n \frac{z^n}{n!} \sum_{k=0}^{\infty} c_k z^k \tag{2.28}$$

with the coefficients

$$c_0 = 2^n,\; c_k = \frac{k - n - 1}{4k} c_{k-1}\,. \tag{2.29}$$

The simple form of the coefficients (2.29) is a result of the fact that only a_0 and a_1 are not equal to zero. In the last step of the series expansion the infinite double sum in (2.28) is changed to one infinite and one finite sum using the Cauchy product of two series (2.16). This results in the power series expansion of function 1

$$\mathrm{e}^{-a\frac{\frac{3}{2} - 2z + \frac{1}{2}z^2}{\Delta t}} = \mathrm{e}^{-a\frac{3}{2\Delta t}} \sum_{n=0}^{\infty} z^n \sum_{k=0}^{n} \left(\frac{a}{\Delta t}\right)^k \frac{1}{k!} c_{n-k} = \sum_{n=0}^{\infty} \omega_n z^n\,, \tag{2.30}$$

with the BDF 2 as the underlying multistep method. In representation (2.30), the integration weights ω_n can easily be identified as the coefficients of the series.

The series expansion of f_2 in (2.24) proceeds in the same way. At first, the Laplace transform of f_2 has to be determined. This function is normally not found in tables, therefore, the definition of the Laplace transform is used. The integral

$$\begin{aligned}\int_0^\infty tH(t-a)\,\mathrm{e}^{-st}\,\mathrm{d}t = \int_a^\infty t\mathrm{e}^{-st}\,\mathrm{d}t &= \left[t\left(-\frac{\mathrm{e}^{-st}}{s}\right)\right]_a^\infty + \int_a^\infty \frac{1}{s}\mathrm{e}^{-st}\,\mathrm{d}t \\ &= \mathrm{e}^{-as}\left(\frac{a}{s} + \frac{1}{s^2}\right)\end{aligned} \tag{2.31}$$

is solved using integration by parts. Now, in the right hand side of (2.31) the Laplace variable is replaced by $s = \gamma(z)/\Delta t$. Additionally to the procedure leading to the analytical series expansion of f_1, here, the division of power series has to be introduced

$$\frac{\sum\limits_{k=0}^{\infty} b_k z^k}{\sum\limits_{k=0}^{\infty} a_k z^k} = \frac{1}{a_0}\sum_{k=0}^{\infty} c_k z^k \quad \text{with} \quad c_k = b_k - \frac{1}{a_0}\sum_{\ell=1}^{k} c_{k-\ell}\,a_\ell\,. \tag{2.32}$$

Then taking (2.32) and the power series expansion of f_1 (2.30) the series expansion of f_2 at $s = \gamma(z)/\Delta t$ is achieved

$$e^{-a\frac{\frac{3}{2}-2z+\frac{1}{2}z^2}{\Delta t}}\left(\frac{a\Delta t}{\frac{3}{2}-2z+\frac{1}{2}z^2}+\frac{\Delta t^2}{\left(\frac{3}{2}-2z+\frac{1}{2}z^2\right)^2}\right)=a^2\sum_{n=0}^{\infty}\left(\omega_n^{(1)}+\omega_n^{(2)}\right)z^n\,, \tag{2.33}$$

where each coefficient $\omega_n^{(i)}$

$$\omega_n^{(1)}=\frac{2}{3}e^{-a\frac{3}{2\Delta t}}\sum_{k=0}^{n}\left(\frac{a}{\Delta t}\right)^{k-1}\frac{c_{n-k}}{k!}+\frac{4}{3}\omega_{n-1}^{(1)}-\frac{1}{3}\omega_{n-2}^{(1)} \tag{2.34a}$$

$$\omega_n^{(2)}=\frac{4}{9}e^{-a\frac{3}{2\Delta t}}\sum_{k=0}^{n}\left(\frac{a}{\Delta t}\right)^{k-2}\frac{c_{n-k}}{k!}+\frac{8}{3}\omega_{n-1}^{(2)}-\frac{22}{9}\omega_{n-2}^{(2)}+\frac{8}{9}\omega_{n-3}^{(2)}-\frac{1}{9}\omega_{n-4}^{(2)} \tag{2.34b}$$

corresponds to one term in the brackets on the left side of equation (2.33). The coefficients c_{n-k} in (2.34) are the same as in (2.29).

2.2.2 Computing the integration weights ω_n

If the convolution quadrature method is applied to solve integral 1 and 2 the necessary integration weights ω_n could be determined either by the series expansion (2.30) and (2.33) or by the numerical evaluation of the integral (2.22). In Fig. 2.2, the integration weights are depicted versus n. Additionally, the error

$$E=\left|\frac{\omega_n^{numerical}-\omega_n^{series}}{\omega_n^{series}}\right| \tag{2.35}$$

in a logarithmic scaling is presented. In definition (2.35), $\omega_n^{numerical}$ denote those weights calculated by formula (2.22) and ω_n^{series} those weights calculated by the series expansion (2.30) or (2.33). The results are very similar so that in Fig. 2.2 only one line is observed. Larger errors occur only for ω_n close to zero. In this case both formulas try to approximate zero, whereas the approximation of $\omega_n^{numerical} \approx 0$ by (2.22) is several decades larger than the approximation $\omega_n^{series} \approx 0$. Anyhow, the "numerical zeros" of the integration weights can be accepted to be a good approximation of zero, e.g., the weights for function 2 are $\omega_n^{numerical} \approx 10^{-6}$ and $\omega_n^{series} \approx 10^{-10}$ for $n < 6$. First conclusions from these small deviations between the different formulas lead to qualify the parameter $L = N$ and $\mathscr{R}^N = \sqrt{\varepsilon}$ used in formula (2.22) as a good choice.

The next interesting question is: How is the influence of L on the results with fixed $\mathscr{R}^N = \sqrt{\varepsilon}$, $\varepsilon = 10^{-1}$? The value of L in equation (2.22) is the amount of steps used in the trapezoidal rule for the approximation of the integration in (2.21). Therefore, varying L corresponds to a study concerning the approximation accuracy of the integral in (2.21). The integration weights $\omega_n^{numerical}$ for both functions (2.23) and

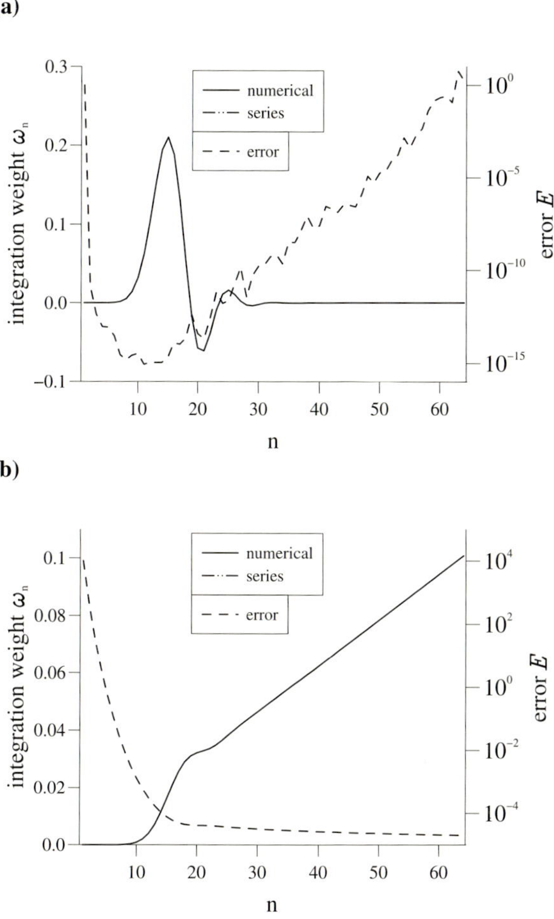

Fig. 2.2. Comparison of ω_n calculated by the numerical approach or by series expansion **(a)** Function f_1 **(b)** Function f_2

(2.24) and some exemplary values of $L = 0.75N, N, 2N$ are plotted in Fig. 2.3 on the left side versus n. On the right hand side of Fig. 2.3 the corresponding errors are given in logarithmic scale versus n. For $L = 0.75N$ at the distinct point $n = 45$ the results become worse, i.e., the error increases rapidly. This point changes to smaller values $n < 45$ if $L < 0.75N$ is decreased and tends to infinity if L approaches $L = N$, i.e., for $L \geq N$ the error is less than $10^{-4} = 0.01\%$ apart from the approximation of the "numerical zeros" as discussed before. Even though the error decreases for $L \geq N$ in function f_2 in the following $L = N$ is used, because either for function f_1 the error does not decrease and a error of 0.01% is sufficient small.

An improvement of the integration accuracy could also be to chose an integration formula with higher order than the trapezoidal rule, e.g., Simpson's rule. How-

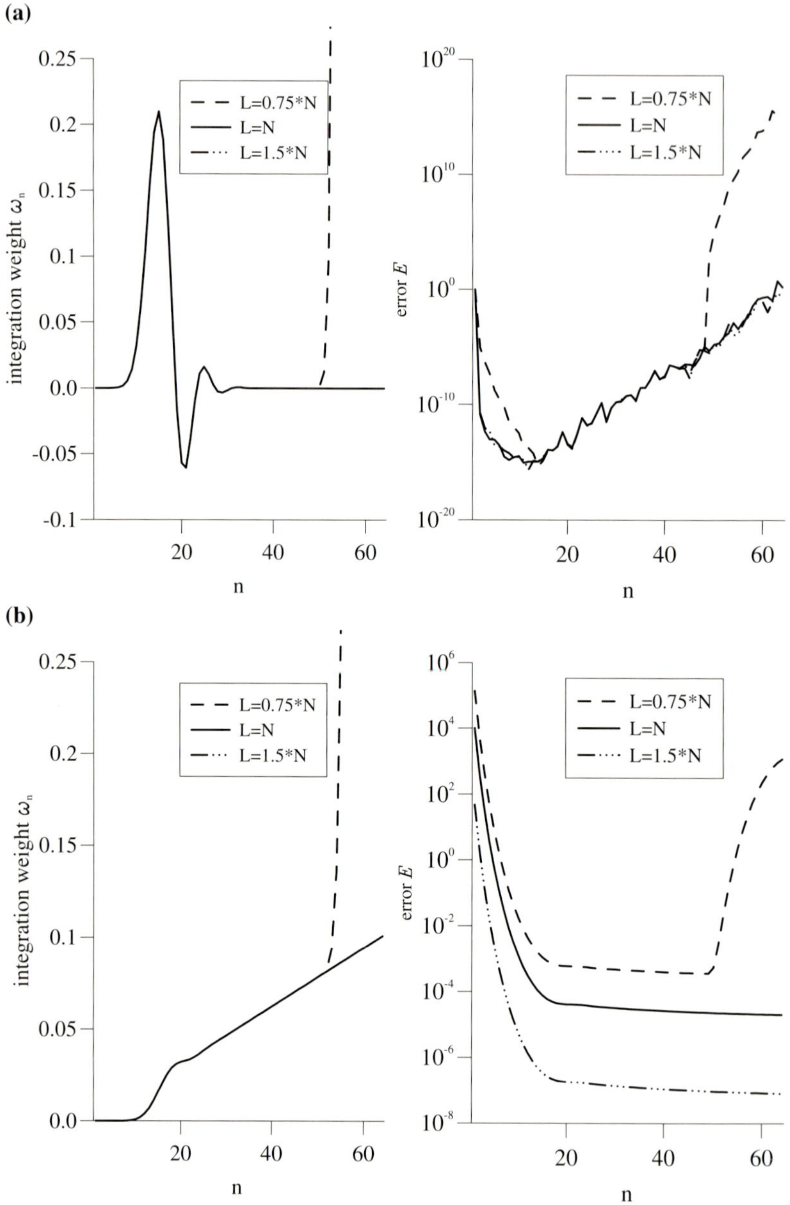

(a)

L=0.75*N
L=N
L=1.5*N
integration weight ω_n
error E
n

(b)

L=0.75*N
L=N
L=1.5*N
integration weight ω_n
error E
n

Fig. 2.3. Influence of L on the integration weights $\omega_n^{numerical}$ **(a)** Function f_1 **(b)** Function f_2

ever, parameter studies of L with the Simpson rule do not lead to higher integration accuracy. Concluding, the trapezoidal rule with $L = N$ in formula (2.22) is the best choice taking accuracy and the numerical effort into account, particularly, because the FFT algorithm is only usable for the calculation of the integration weights for $L = N$.

Now, the influence of ε, the assumed computing error of the function $\hat{f}(s)$ (see Sect. 2.1), is studied. This error is normally not known and can only be guessed. In

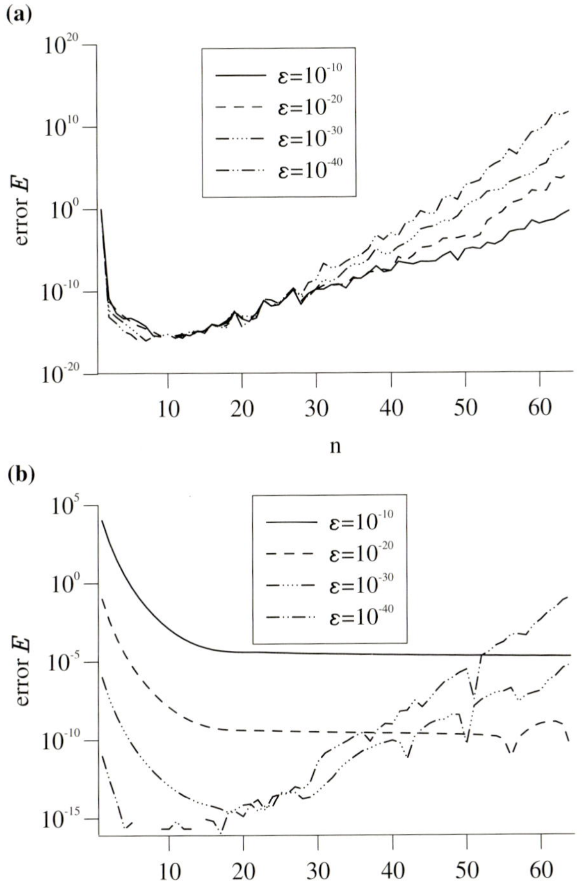

Fig. 2.4. Influence of ε on the accuracy of the integration **(a)** Function f_1 **(b)** Function f_2

Fig. 2.4, the error E (2.35) is plotted in logarithmic scaling versus n for different values of $\varepsilon = 10^{-10} \ldots 10^{-40}$. After the first steps independent of ε the error is insignificant. However, the error gets smaller for smaller ε. Surprisingly, for $\varepsilon < 10^{-20}$ the error increases even above the values of $\varepsilon = 10^{-10}$ for large n, and $\varepsilon < 10^{-40}$

lead to completely unsatisfactory results. An explanation for this effect is found, recognizing that such small values of ε yield very small $\mathscr{R}$. This radius $\mathscr{R}$ is the integration path around $z = 0$ in Cauchy's integration formula. Reflecting that the integrand in (2.15) has a singularity at $z = 0$ it is obvious that a too small radius must give worse results. In summary, a value of $\varepsilon = 10^{-10}$ or $\varepsilon = 10^{-20}$ is preferable. Here, in the following $\varepsilon = 10^{-10}$ will be used.

Finally, it can be stated that the integration in (2.22) with $L = N$ and $\mathscr{R}^N = \sqrt{\varepsilon}$, $\varepsilon = 10^{-10}$ yields as good results as the analytical development of the power series (2.14).

2.2.3 Numerical convolution

Until now only the calculation of the integration weights ω_n has been investigated. In the following, the convolution integrals 1 and 2 are performed using the results from the last section, i.e., the parameters are chosen: $L = N$ and $\varepsilon = 10^{-10}$. Figure 2.5 shows the numerical results achieved by the convolution quadrature method as well as their exact solutions using the values $a = 0.5$ and $b = 2$ (see (2.23) and (2.24)). The numerical results and the exact solution are in good agreement and even the time step size Δt has nearly no influence if it is small enough. For $\Delta t = 0.15\,\mathrm{s}$ the step size is obviously to coarse in case of integral 1 having in mind that after 3 time steps the first jump has to be approximated there. For integral 2 even this coarse time approximation gives sufficient results. The jumps in integral 1 are approximated very good for the smaller time step sizes, however, a overshooting comparable to Gibbs phenomenon occur. The overshooting has smaller amplitudes but a wider spread for large time step sizes and higher amplitudes with a more located spread for small time step sizes.

To obtain the results in Fig. 2.5 the numerical approximation of the integration weights ω_n (2.22) is used. As expected from the last tests, using instead the series expansion of ω_n (2.30) or (2.33) similar results are achieved with the exception for $\Delta t = 0.01\,\mathrm{s}$. With this time step size no reasonable results using the series expansion are achieved, probably due to rounding errors. The numerical effort for both methods measured in CPU-time is the same.

Another interesting question is about the underlying multistep method. The mathematical proofs in [118, 121, 120] prescribe A-stable or for some functions $A(\alpha)$-stable methods which must be also stable in infinity. In the following, a BDF 1, BDF 3 and the trapezoidal rule are compared to the BDF 2 used in all tests before. The BDF 1, BDF 2, and the trapezoidal rule are A-stable whereas the BDF 3 is only $A(\alpha)$-stable. In Fig. 2.6, integral 1 is computed with the four mentioned multistep methods. For the BDF 1 a time step size $\Delta t = 0.005\,\mathrm{s}$ and for the other three methods $\Delta t = 0.04\,\mathrm{s}$ is used. The results for integral 2 are not presented because in this case all four tested multistep methods give the same good result. Contrary, the BDF 3 and the trapezoidal rule are ineligible to calculate integral 1 as clearly observed in Fig. 2.6. The oscillating after the jumps does not stop using the trapezoidal rule and it is too large for the BDF 3. These results confirm the mathematics which enforce stability at infinity which is not fulfilled by the trapezoidal rule. In the same manner,

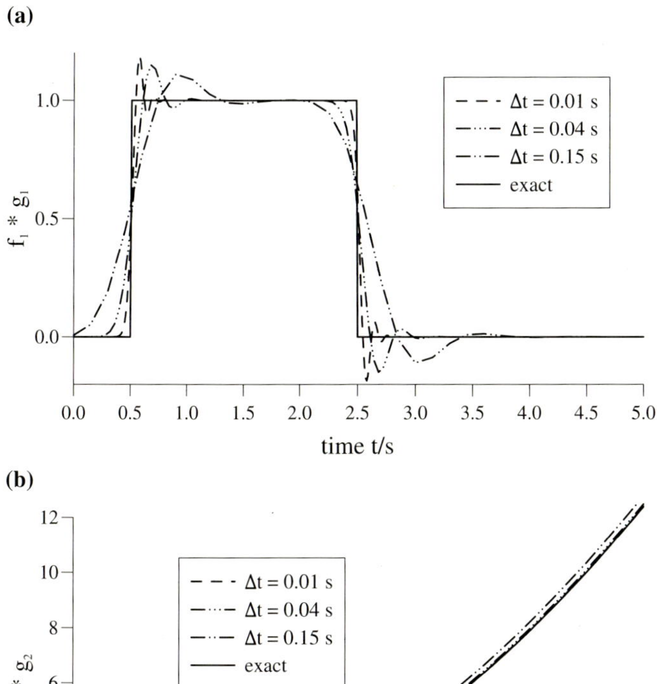

Fig. 2.5. Convolution Quadrature method using different time step sizes Δt **(a)** Integral 1 **(b)** Integral 2

$A(\alpha)$-stable methods are shown to be not sufficient if the integral results in functions with jumps, e.g., in wave propagation problems, as demonstrated by the results using the BDF 3. The A-stable methods BDF 1 and the BDF 2 lead to good results, but the BDF 1 needs very small time step sizes (factor 10) compared to the BDF 2. Contrary to the BDF 2, there is no overshooting at the jumps. Consequently, the BDF 2 seems to be the best choice if efficiency and accuracy is taken into account.

In practical applications, the convolution integral (2.1) is often solved numerically using the property of the Laplace transform

$$f * g = \int_0^t f(t-\tau)\, g(\tau)\, \mathrm{d}\tau \qquad \circ\!\!-\!\!\bullet \qquad \hat{f}(s)\, \hat{g}(s) \,, \tag{2.36}$$

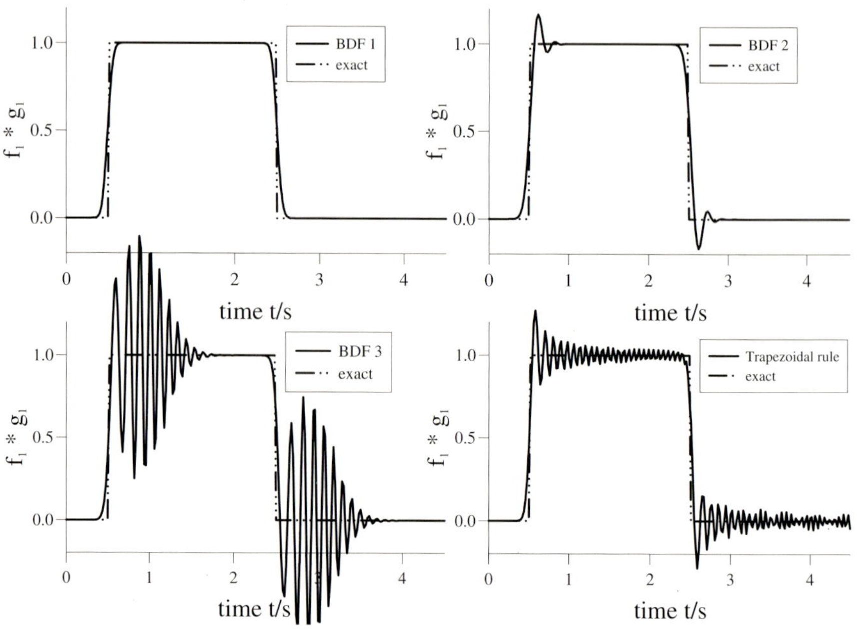

Fig. 2.6. Convolution quadrature using different multistep methods

with a subsequent numerical inverse transformation. If, however, the Laplace transform of either f or g does not exist performing the convolution integral via the Laplace domain is not possible. In case of piecewise defined boundary conditions appearing, e.g., in contact problems, no Laplace transform exist. In such cases the convolution quadrature method has its biggest advantage, which is the ability to evaluate a convolution integral without the knowledge of either $\hat{f}(s)$ or $\hat{g}(s)$.

But also, if the Laplace transform of both functions f and g are available the convolution quadrature method can be used as an inverse Laplace transformation method. Therefore, in the following a comparison with some inverse transformation methods is presented. A lot of methods are available in the literature for this task (see, e.g., [53]). In [133] several methods are tested keeping boundary value problems in mind. In the final conclusion the method of Dubner and Abate [78] is favored. Two variants of this method from Durbin [79] and Crump [58] will be compared in the following with the convolution quadrature method. In Fig. 2.7, the results of the two mentioned methods, the convolution quadrature method, abbreviated by CQM, and the exact solution are depicted versus time. None of the methods give worse results, whereas the methods of Durbin and Crump lead to nearly equal results, and, therefore, are not distinguishable in the plot. The convolution quadrature method is superior reproducing the jump at the wave fronts. However, the disadvantage of the convolution quadrature method is the overshooting after the jump. But more important, for getting the results presented in Fig. 2.7 the inverse transformation methods from Durbin as well as from Crump need two parameters to

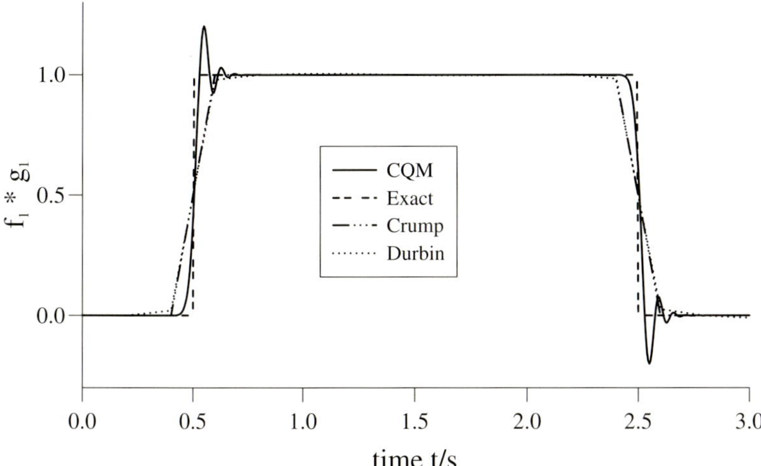

Fig. 2.7. Convolution quadrature used as an inverse Laplace transform: Comparison with Durbin's and Crump's method

be adjusted. A wrong choice of these parameters lead to completely unsatisfactory results. Also, these parameters can only for smooth functions chosen with standard values. In the presented case, however, several trail and error steps were necessary. Instead for the convolution quadrature method only the time step size has to be chosen small enough. This is not to say that the convolution quadrature method is in general the better inverse transformation method, but for functions with jumps as appearing in wave propagation problems the convolution quadrature method is the better choice for early times. In the long time behavior there are probably better methods available, because a numerical damping occur when applying the convolution quadrature method.

3. Viscoelastically supported Euler-Bernoulli beam

A field of application for the convolution quadrature method are time dependent integral equations. Here, the integral equation for a transient excited viscoelastically supported Euler-Bernoulli beam will be deduced and solved with the convolution quadrature method. A direct evaluation in time domain is only possible without the viscoelastic foundation, however, the resulting time stepping procedure is not stable [67]. Also, the time-dependent fundamental solutions for the beam theory needed to solve directly in time domain contain Fresnel integrals. Both points, unstable time stepping procedure and complicated time-dependent fundamental solutions, give reason to use the convolution quadrature method. Moreover, except by means of the convolution quadrature method, for the beam on a viscoelastic foundation there is no possibility to establish a formulation in time domain since no time-dependent fundamental solution is known.

3.1 Integral equation for a beam resting on viscoelastic foundation

Consider a thin beam of length ℓ on a viscoelastic foundation undergoing transverse motion $w(x)$ caused by a load per length $q(x,t)$. As depicted in Fig. 3.1, the longitudinal coordinate is x, the flexural rigidity EI, the density ρ, and the beam cross section A. The foundation is described by the spring constant c and the damping

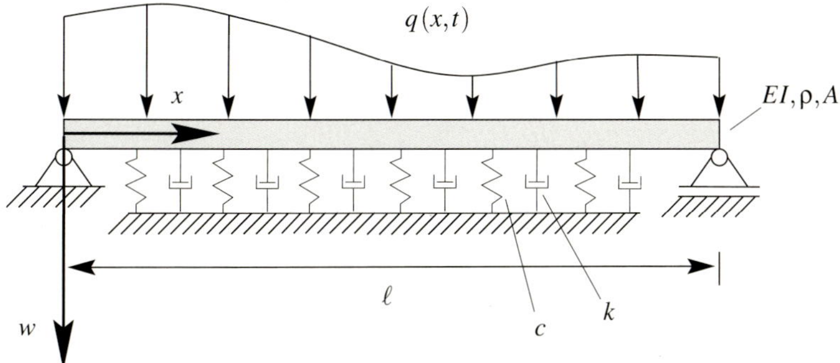

Fig. 3.1. Simply supported beam on a viscoelastic foundation

coefficient k. The basic hypothesis of the Euler-Bernoulli theory of beams leads to the governing partial differential equation [99]

$$EIw^{(IV)}(x,t) + cw(x,t) + k\dot{w}(x,t) + \rho A\ddot{w}(x,t) = q(x,t) \;, \tag{3.1}$$

where the partial differentiation with respect to x is denoted by the superscript $^{(IV)}$ and with respect to time t by the over-dot. In the following, it is assumed that the initial conditions for $w(x,t=0)=0$ and $\dot{w}(x,t=0)=0$ vanish. The corresponding integral equation is achieved with the weighted residual technique

$$\begin{aligned}\int_0^t\int_0^\ell \left[EIw^{(IV)}(x,\tau) + cw(x,\tau) + k\dot{w}(x,\tau) + \rho A\ddot{w}(x,\tau)\right] w^\star(x,y,t-\tau)\,\mathrm{d}x\mathrm{d}\tau \\ = \int_0^t\int_0^\ell q(x,\tau)\, w^\star(x,y,t-\tau)\,\mathrm{d}x\mathrm{d}\tau\end{aligned} \tag{3.2}$$

using the time-dependent fundamental solution of the deflection $w^\star(x,y,t-\tau)$. For the next step, this solution should be known. As mentioned in the introduction, here, the convolution quadrature method will be used because no time domain fundamental solution is known. Hence, the deduction of the integral equation is performed in Laplace domain. The Laplace transform of the integral equation (3.2) yields

$$\int_0^\ell EI\left[\hat{w}^{(IV)}(x) + \kappa^4\hat{w}(x)\right]\hat{w}^\star(x,y)\,\mathrm{d}x = \int_0^\ell \hat{q}(x)\,\hat{w}^\star(x,y)\,\mathrm{d}x \;, \tag{3.3}$$

with the abbreviation $\kappa^4 = \left(c + ks + \rho As^2\right)/EI$. For the following, the Laplace transformed fundamental solution is required.

3.1.1 Fundamental solutions

In the literature, several fundamental solutions for the Euler-Bernoulli beam are given, e.g., in [9, 67, 48, 146]. The latter, [146], is a special fundamental solution with extra terms. These terms do not influence the singularity but enforce zero deflection and zero bending moment at distance ℓ from the singularity. From a mathematical point of view, these extra terms are not necessary, but they have some practical advantages when solving a bending vibration problem of a beam with length ℓ.

The fundamental solution given in [67] can be deduced with the methods presented in [48] and can be written for the deflection as ($\kappa' = \kappa/\sqrt{2}$)

$$\hat{w}_L^\star(x,y) = \frac{1}{8EI\,\kappa'^3}\mathrm{e}^{-\kappa'|x-y|}\left(\cos\left(\kappa'|x-y|\right) + \sin\left(\kappa'|x-y|\right)\right) \;. \tag{3.4}$$

According to the definition in [9], the completeness of a fundamental solution is given if the special case $\kappa = 0$, i.e., the static case, is included in the fundamental solution (for details see [9]). In the same paper, it was proven that the above fundamental solution (3.4) is not complete, whereas the following function ($\lambda = \sqrt{\mathrm{i}}\,\kappa$)

$$\hat{w}^{\star}(x,y) = \frac{1}{4EI\,\lambda^3}\left[\sinh\left(\lambda|x-y|\right) - \sin\left(\lambda|x-y|\right)\right] \tag{3.5}$$

is a complete fundamental solution in Laplace domain.

To verify the completeness, i.e., to show whether the static case is included or not, the limit $\kappa' \to 0$ in equation (3.4) and $\lambda \to 0$ in equation (3.5), respectively, has to be examined. The infinite power series of the functions $\mathrm{e}^{\lambda r}$, $\sin(\lambda r)$, $\cos(\lambda r)$, and $\sinh(\lambda r)$ with respect to $\lambda r = \lambda|x-y|$ are

$$\begin{aligned}
\mathrm{e}^{\lambda r} &= 1 + \lambda r + \frac{(\lambda r)^2}{2!} + \frac{(\lambda r)^3}{3!} + \mathcal{O}\left((\lambda r)^4\right)\\
\sin(\lambda r) &= \lambda r - \frac{(\lambda r)^3}{3!} + \mathcal{O}\left((\lambda r)^5\right)\\
\cos(\lambda r) &= 1 - \frac{(\lambda r)^2}{2!} + \mathcal{O}\left((\lambda r)^4\right)\\
\sinh(\lambda r) &= \lambda r + \frac{(\lambda r)^3}{3!} + \mathcal{O}\left((\lambda r)^5\right)
\end{aligned}$$

leading to the infinite series representation of the fundamental solutions (3.4) and (3.5)

$$\begin{aligned}
\hat{w}_L^{\star}(x,y) =& \frac{1}{8EI\,\kappa'^3}\left[\left(1 - \kappa' r + \frac{(\kappa' r)^2}{2} - \frac{(\kappa' r)^3}{6}\right)\right.\\
&\left.\left(1 + \kappa' r - \frac{(\kappa' r)^2}{2} - \frac{(\kappa' r)^3}{6}\right) + \mathcal{O}\left((\kappa' r)^4\right)\right]\\
=& \frac{1}{8EI\,\kappa'^3}\left[\left(1 - (\kappa' r)^2 + \frac{2}{3}(\kappa' r)^3\right) + \mathcal{O}\left((\kappa' r)^4\right)\right]\\
\hat{w}^{\star}(x,y) =& \frac{1}{4EI\,\lambda^3}\left[\left(\lambda r + \frac{(\lambda r)^3}{3!}\right) - \left(\lambda r - \frac{(\lambda r)^3}{3!}\right) + \mathcal{O}\left((\lambda r)^5\right)\right]\\
=& \frac{1}{4EI\,\lambda^3}\left[2\,\frac{(\lambda r)^3}{6} + \mathcal{O}\left((\lambda r)^5\right)\right].
\end{aligned} \tag{3.6}$$

Now, in the series expansions (3.6) the limit $\lambda \to 0$ or $\kappa' \to 0$, respectively, can be performed yielding either the static fundamental solution or an infinite value

$$\lim_{\lambda\to 0}\hat{w}^{\star}(x,y) = \frac{r^3}{12EI} = w_{static}(x,y) \quad \text{or} \quad \lim_{\kappa'\to 0}\hat{w}_L^{\star}(x,y) = \infty\,. \tag{3.7}$$

This limit shows that the fundamental solution (3.5) is complete contrary to the solution (3.4).

Another condition which most fundamental solution fulfill is the Sommerfeld radiation condition, e.g., in 3-d elastodynamics this condition is fulfilled [84] (for a detailed discussion see [148]). Simplifying, the Sommerfeld radiation condition ensures two properties: first, only outward propagation no inward propagation of waves is possible, i.e., there is no wave reflection at infinity, and, second, all waves decay with distance from the excitation point, i.e., they tend to zero for increasing distance r. These physical conditions are checked for the beam fundamental solutions. By the definition of the Laplace transform (see appendix A.3) $\Re(s) > 0$ is given yielding $\Re(\kappa') > 0$ but no prediction for λ. However, the equivalence

$$\sinh(\lambda r) = \frac{1}{2}\left(\mathrm{e}^{\lambda r} - \mathrm{e}^{-\lambda r}\right) \tag{3.8}$$

shows that either one of the exponential functions in equation (3.8) tend to infinity for increasing r whatever sign the real part of λ has. Contrary, the exponential function in the fundamental solution (3.4) with the condition of a positive real part of κ' ensures a decreasing value for increasing r. In summary, the limits

$$\lim_{r\to\infty} \hat{w}^\star(x,y) = \infty \quad \text{and} \quad \lim_{r\to\infty} \hat{w}_L^\star(x,y) = 0 \tag{3.9}$$

hold.

The above shows that either the static case can be included in the fundamental solution or the radiation condition is fulfilled but not both. Because in the following no infinite beams are considered, the fundamental solution (3.5) will be used. In summary with all needed derivatives it is:

$$\hat{w}^\star(x,y) = \frac{1}{4EI\lambda^3}\left[\sinh(\lambda|x-y|) - \sin(\lambda|x-y|)\right] \tag{3.10a}$$

$$\begin{aligned}\hat{w'}^\star(x,y) &= \frac{\partial}{\partial x}\hat{w}^\star(x,y)\\ &= \frac{1}{4EI\lambda^2}\left[\cosh(\lambda|x-y|) - \cos(\lambda|x-y|)\right](2H(x-y)-1)\end{aligned} \tag{3.10b}$$

$$\begin{aligned}\hat{M}^\star(x,y) &= -EI\frac{\partial^2}{\partial x^2}\hat{w}^\star(x,y)\\ &= \frac{-1}{4\lambda}\left[\sinh(\lambda|x-y|) + \sin(\lambda|x-y|)\right]\end{aligned} \tag{3.10c}$$

$$\begin{aligned}\hat{Q}^\star(x,y) &= -EI\frac{\partial^3}{\partial x^3}\hat{w}^\star(x,y)\\ &= \frac{1}{4}\left[\cosh(\lambda|x-y|) + \cos(\lambda|x-y|)\right](1-2H(x-y))\end{aligned} \tag{3.10d}$$

with the fundamental solution of the bending moment $\hat{M}^\star(x,y)$ and the shear force $\hat{Q}^\star(x,y)$. The fundamental solution $\hat{w}^\star(x,y)$ depends only on the absolute distance

$r = |x-y|$ of the load point y and field point x, which represents the physical fact that the deflection is independent whether the impact is on y and the observation on x or vice versa. The mathematical consequence is a continuous but not differentiable function $\hat{w}^\star(x,y)$ at the point $x = y$. Due to this, the first derivative of $\hat{w}^\star(x,y)$ and the shear force $\hat{Q}^\star(x,y)$ has a jump at $x = y$, denoted by $(2H(x-y)-1)$. The fundamental solution $\hat{w}^\star(x,y)$ is the response of the beam caused by a single force at y. Therefore, the shear force must jump at $x = y$ to fulfill the mechanical equilibrium of forces.

Finally, it should be remarked that all results presented in the following are also achievable with the other fundamental solution (3.4).

3.1.2 Integral equation

Now, an integration by parts of the first term in (3.3) will be performed. Because the point $x = y$ is not differentiable, i.e., also not integrable, the integral is divided in two integrable parts

$$\begin{aligned}
&\int_0^\ell EI\hat{w}^{(IV)}(x)\,\hat{w}^\star(x,y)\,\mathrm{d}x \\
&\quad = \lim_{\varepsilon\to 0}\left(\int_0^{y-\varepsilon} EI\hat{w}^{(IV)}(x)\,\hat{w}^\star(x,y)\,\mathrm{d}x + \int_{y+\varepsilon}^{\ell} EI\hat{w}^{(IV)}(x)\,\hat{w}^\star(x,y)\,\mathrm{d}x\right)
\end{aligned} \tag{3.11}$$

with a subsequent limiting process. Proceeding in this way, after four partial integrations the final form is reached. Only the last partial integration is explained in detail in the following due to its crucial result. After three partial integrations it is found

$$\begin{aligned}
&\int_0^\ell EI\hat{w}^{(IV)}(x)\,\hat{w}^\star(x,y)\,\mathrm{d}x \\
&\quad = \left[-\hat{Q}(x)\,\hat{w}^\star(x,y) + \hat{M}(x)\,\hat{w}'^\star(x,y) - \hat{w}'(x)\,\hat{M}^\star(x,y)\right]_{x=0}^{\ell} \\
&\qquad + \lim_{\varepsilon\to 0}\left(\int_0^{y-\varepsilon} \frac{\hat{w}'(x)}{4}\left[\cosh(\lambda(y-x)) + \cos(\lambda(y-x))\right]\mathrm{d}x\right. \\
&\qquad\qquad \left. - \int_{y+\varepsilon}^{\ell} \frac{\hat{w}'(x)}{4}\left[\cosh(\lambda(x-y)) + \cos(\lambda(x-y))\right]\mathrm{d}x\right)
\end{aligned} \tag{3.12}$$

using the corresponding expressions to (3.10c) and (3.10d) for the bending moment $\hat{M}(x)$ and the shear force $\hat{Q}(x)$, respectively. In every step, the terms concerning the points $y \pm \varepsilon$ cancel each other in the limit $\varepsilon \to 0$. In contrary, in the remaining integral of (3.12) this limiting process gives an additional term. In detail it is

$$\begin{aligned}
&\lim_{\varepsilon\to 0}\left(\int_0^{y-\varepsilon} \frac{\hat{w}'(x)}{4}\left[\cosh\left(\lambda\left(y-x\right)\right)+\cos\left(\lambda\left(y-x\right)\right)\right]\mathrm{d}x\right.\\
&\qquad\qquad \left. -\int_{y+\varepsilon}^{\ell} \frac{\hat{w}'(x)}{4}\left[\cosh\left(\lambda\left(x-y\right)\right)+\cos\left(\lambda\left(x-y\right)\right)\right]\mathrm{d}x\right)\\
&=\left[\frac{\hat{w}(x)}{4}\left[\cosh\left(\lambda|x-y|\right)+\cos\left(\lambda|x-y|\right)\right]\left(1-2H\left(x-y\right)\right)\right]_{x=0}^{\ell}\\
&+\lim_{\varepsilon\to 0}\left[\frac{\hat{w}(x)}{4}\left[\cosh\left(\lambda|x-y|\right)+\cos\left(\lambda|x-y|\right)\right]\left(2H\left(x-y\right)-1\right)\right]_{y-\varepsilon}^{y+\varepsilon}\\
&+\int_0^{\ell}\frac{\hat{w}(x)\,\lambda}{4}\left[\sinh\left(\lambda|x-y|\right)-\sin\left(\lambda|x-y|\right)\right]\mathrm{d}x\,.
\end{aligned} \tag{3.13}$$

The first term on the right hand side is the product of the deflection $\hat{w}$ with the fundamental solution of the shear force $\hat{Q}^{\star}(x,y)$ (3.10d). In the next term, the limiting process leads to the deflection $\hat{w}$ itself

$$\begin{aligned}
&\lim_{\varepsilon\to 0}\left[\frac{\hat{w}(x)}{4}\left[\cosh\left(\lambda|x-y|\right)+\cos\left(\lambda|x-y|\right)\right]\left(2H\left(x-y\right)-1\right)\right]_{y-\varepsilon}^{y+\varepsilon}\\
&=\lim_{\varepsilon\to 0}\frac{\cosh\left(\lambda\varepsilon\right)+\cos\left(\lambda\varepsilon\right)}{4}\left[\hat{w}\left(y+\varepsilon\right)\left(2H\left(\varepsilon\right)-1\right)-\hat{w}\left(y-\varepsilon\right)\left(2H\left(-\varepsilon\right)-1\right)\right]\\
&=\hat{w}\left(y\right)\,.
\end{aligned} \tag{3.14}$$

The last term in equation (3.13) is identified as ($\lambda^4=\left(\sqrt{\mathrm{i}}\kappa\right)^4=-\kappa^4$)

$$\begin{aligned}
\int_0^{\ell} EI\hat{w}(x)\,\lambda\frac{\sinh\left(\lambda|x-y|\right)-\sin\left(\lambda|x-y|\right)}{4EI}\mathrm{d}x &= \int_0^{\ell} EI\hat{w}(x)\,\lambda^4\hat{w}^{\star}(x,y)\,\mathrm{d}x\\
&=-\int_0^{\ell} EI\hat{w}(x)\,\kappa^4\hat{w}^{\star}(x,y)\,\mathrm{d}x
\end{aligned} \tag{3.15}$$

which is equal to the second part of the integral on the left hand side of equation (3.3). Due to the negative sign in front of the integral in (3.15) these two terms cancel each other. With these considerations the integral equation is reduced to the equation

$$\hat{w}(y) + \Big[-\hat{Q}(x)\,\hat{w}^{\star}(x,y) + \hat{M}(x)\,\hat{w'}^{\star}(x,y) \\ - \hat{w}'(x)\,\hat{M}^{\star}(x,y) + \hat{w}(x)\,\hat{Q}^{\star}(x,y) \Big]_{x=0}^{\ell} = \int_0^{\ell} \hat{q}(x)\,\hat{w}^{\star}(x,y)\,\mathrm{d}x \tag{3.16}$$

with unknowns only on the "boundary", i.e., at the end points of the beam. Finally, a formal inverse Laplace transform leads to the integral equation in time domain

$$w(y,t) + \big[-Q(x,t) * w^{\star}(x,y,t) + M(x,t) * w'^{\star}(x,y,t) - w'(x,t) * M^{\star}(x,y,t) \\ + w(x,t) * Q^{\star}(x,y,t) \big]_{x=0}^{\ell} = \int_0^{\ell} q(x,t) * w^{\star}(x,y,t)\,\mathrm{d}x\,. \tag{3.17}$$

The inverse transformation in equation (3.17) was called formal because, as mentioned above, the necessary time-dependent fundamental solutions are not known. Therefore, here, the convolutions quadrature method has an ideal application. Applying the quadrature formula (2.18) to the convolution integrals in (3.17) a time stepping procedure for $w(y,t)$ and $n = 0,1,\ldots,N$ is achieved

$$w(y,n\Delta t) + \sum_{k=0}^{n} \Big[-\omega_{n-k}(\hat{w}^{\star},x,y,\Delta t)\,Q(x,k\Delta t) + \omega_{n-k}\left(\hat{w'}^{\star},x,y,\Delta t\right) M(x,k\Delta t) \\ - \omega_{n-k}\left(\hat{M}^{\star},x,y,\Delta t\right) w'(x,k\Delta t) + \omega_{n-k}\left(\hat{Q}^{\star},x,y,\Delta t\right) w(x,k\Delta t) \Big]_{x=0}^{\ell} \\ = \int_0^{\ell} \sum_{k=0}^{n} \omega_{n-k}(\hat{w}^{\star},x,y,\Delta t)\,q(x,k\Delta t)\,\mathrm{d}x\,, \tag{3.18}$$

using only the Laplace transformed fundamental solutions (3.10), i.e., the integration weights are determined by equation (2.22) using for $\hat{f}$ the appropriate fundamental solution as indicated in the argument list, e.g.,

$$\omega_{n-k}(\hat{w}^{\star},x,y,\Delta t) = \frac{\mathscr{R}^{-(n-k)}}{L} \sum_{\ell=0}^{L-1} \hat{w}^{\star} \left(\frac{\gamma\left(\mathscr{R}\mathrm{e}^{\mathrm{i}\ell\frac{2\pi}{L}}\right)}{\Delta t} \right) \mathrm{e}^{-\mathrm{i}(n-k)\ell\frac{2\pi}{L}}\,. \tag{3.19}$$

With the integral equation (3.18) the deflection of a beam at every point y and time t can be calculated if all boundary data, the deflection, the slope, the bending moment, and the shear force at both ends of the beam are known. But, for every well posed problem only half of the data are given by the boundary conditions. The data corresponding to the boundary condition, respectively, are unknown, e.g., if the deflection is given the shear forces are unknown. Due to this, as usual in boundary element methods, first the unknown boundary data are determined taking the point y to the boundary, i.e., for a beam to the end points $y = 0,\ell$. Then, with equation

(3.18) at both points two equations for four unknowns are available. Therefore, a second integral equation is necessary. Mostly, the integral equation for the slope $\varphi(y) = -\partial w(y)/\partial y$ is used. The weighted residual statement is for this case

$$\int_0^t \int_0^\ell \left[EIw^{(IV)}(x,\tau) + cw(x,\tau) + k\dot{w}(x,\tau) + \rho A\ddot{w}(x,\tau)\right]\varphi^\star(x,y,t-\tau)\,\mathrm{d}x\mathrm{d}\tau$$
$$= \int_0^t \int_0^\ell q(x,\tau)\,\varphi^\star(x,y,t-\tau)\,\mathrm{d}x\mathrm{d}\tau \tag{3.20}$$

with the fundamental solution $\varphi^\star(x,y,t-\tau) = -\partial w^\star(x,y,t-\tau)/\partial y$. The same partial integration as before for the deflection is performed in Laplace domain. The fundamental solutions (3.10b) - (3.10d) can be used due to the property $\partial w^\star(x,y,t)/\partial y = -\partial w^\star(x,y,t)/\partial x$. Only for the fourth partial integration the fundamental solution

$$-\frac{\partial}{\partial y}\hat{Q}^\star(x,y) = \frac{\partial}{\partial x}\hat{Q}^\star(x,y) = \frac{\lambda}{4}\left[\sinh(\lambda|x-y|) - \sin(\lambda|x-y|)\right] \tag{3.21}$$

has to be determined. The integral ought to be divided into two regular parts because at $x = y$ the fundamental solutions are not integrable. Also as before, in the limit the expressions for $y \pm \varepsilon$ in every step cancel each other with one exception. In the second step, the jump in the fundamental solution of the shear force leads to

$$\begin{aligned}
&\lim_{\varepsilon\to 0}\left[\frac{\hat{w}'(x)}{4}\left[\cosh(\lambda|x-y|) + \cos(\lambda|x-y|)\right](2H(x-y) - 1)\right]_{y-\varepsilon}^{y+\varepsilon} \\
&= \lim_{\varepsilon\to 0}\frac{\cosh(\lambda\varepsilon) + \cos(\lambda\varepsilon)}{4}\left[\hat{w}'(y+\varepsilon)(2H(\varepsilon) - 1) - \hat{w}'(y-\varepsilon)(2H(-\varepsilon) - 1)\right] \\
&= \hat{w}'(y)\ .
\end{aligned} \tag{3.22}$$

Finally, an integral equation for the slope in time domain is given

$$\begin{aligned}
-w'(y,t) + \Big[-Q(x,t) * \varphi^\star(x,y,t) + M(x,t) * \varphi'^\star(x,y,t) - w'(x,t) * Q^\star(x,y,t) \\
+ w(x,t) * \frac{\partial}{\partial x}Q^\star(x,y,t)\Big]_{x=0}^{\ell} = \int_0^\ell q(x,t) * \varphi^\star(x,y,t)\,\mathrm{d}x\ .
\end{aligned} \tag{3.23}$$

For consistency with the integral equation (3.17), where the first derivative of the deflection $w'(y,t)$ is used instead of the slope φ, also in the following the derivative of the deflection $w'(y,t)$ is used. These two differ only due to the minus sign in the definition of the slope. Using the convolution quadrature method (2.18) a time stepping procedure for $w'(y,t)$ and $n = 0,1,\ldots,N$ is achieved

$$-w'(y,n\Delta t)+\sum_{k=0}^{n}\Big[-\omega_{n-k}\left(\hat{\varphi}^\star,x,y,\Delta t\right)Q(x,k\Delta t)+\omega_{n-k}\left(\hat{\varphi'}^\star,x,y,\Delta t\right)M(x,k\Delta t)$$
$$-\,\omega_{n-k}\left(\hat{Q}^\star,x,y,\Delta t\right)w'(x,k\Delta t)+\omega_{n-k}\left(\frac{\partial}{\partial x}\hat{Q}^\star,x,y,\Delta t\right)w(x,k\Delta t)\Big]_{x=0}^{\ell}$$
$$=\int_0^\ell\sum_{k=0}^{n}\omega_{n-k}\left(\hat{\varphi}^\star,x,y,\Delta t\right)q(x,k\Delta t)\,\mathrm{d}x\,. \tag{3.24}$$

As mentioned above, both equations (3.24) and (3.18) will be evaluated at the end points of the beam $y=0$ and $y=\ell$ to determine the unknown boundary data, i.e., the fundamental solutions (3.10) and (3.21) are calculated for $r=|x-y|=\ell$ or $r=0$. For $r=0$, i.e., the points x and y coincide, the fundamental solutions are either not integrable but continuous or have a jump as the fundamental solution of the shear forces. Only the jump needs a further consideration. The fundamental solution for the shear forces (3.10d) is in the limit $y\to x$

$$\lim_{y\to x}\hat{Q}^\star(x,y)=\lim_{y\to x}\frac{\cosh(\lambda|x-y|)+\cos(\lambda|x-y|)}{4}\left(1-2H(x-y)\right)$$
$$=\begin{cases}-\frac{1}{2} & y\to x=\ell,\ y<x\\ \frac{1}{2} & y\to x=0,\ y>x\end{cases}, \tag{3.25}$$

because for both ends the limit is taken from the inner point $y>x$ for $x=0$ or $y<x$ for $x=\ell$, respectively. With theses considerations four equations

$$\frac{w(0,n\Delta t)}{2}-\sum_{k=0}^{n}\left[\omega_{n-k}\left(\hat{w}^\star,x,0\right)Q(x,k\Delta t)+\omega_{n-k}\left(\hat{M}^\star,x,0\right)w'(x,k\Delta t)\right]_{x=0}^{\ell}$$
$$+\sum_{k=0}^{n}\omega_{n-k}\left(\hat{w'}^\star,\ell,0\right)M(\ell,k\Delta t)+\sum_{k=0}^{n}\omega_{n-k}\left(\hat{Q}^\star,\ell,0\right)w(\ell,k\Delta t)$$
$$=\int_0^\ell\sum_{k=0}^{n}\omega_{n-k}\left(\hat{w}^\star,x,0\right)q(x,k\Delta t)\,\mathrm{d}x \tag{3.26a}$$

$$\frac{w(\ell,n\Delta t)}{2}-\sum_{k=0}^{n}\left[\omega_{n-k}\left(\hat{w}^\star,x,\ell\right)Q(x,k\Delta t)+\omega_{n-k}\left(\hat{M}^\star,x,\ell\right)w'(x,k\Delta t)\right]_{x=0}^{\ell}$$
$$-\sum_{k=0}^{n}\omega_{n-k}\left(\hat{w'}^\star,0,\ell\right)M(0,k\Delta t)-\sum_{k=0}^{n}\omega_{n-k}\left(\hat{Q}^\star,0,\ell\right)w(0,k\Delta t)$$
$$=\int_0^\ell\sum_{k=0}^{n}\omega_{n-k}\left(\hat{w}^\star,x,\ell\right)q(x,k\Delta t)\,\mathrm{d}x \tag{3.26b}$$

$$\frac{w'(0,n\Delta t)}{2} - \sum_{k=0}^{n}\left[\omega_{n-k}\left(\hat{\varphi}'^{\star},x,0\right)M(x,k\Delta t) + \omega_{n-k}\left(\frac{\partial\hat{Q}^{\star}}{\partial x},x,0\right)w(x,k\Delta t)\right]_{x=0}^{\ell}$$
$$+\sum_{k=0}^{n}\omega_{n-k}\left(\hat{\varphi}^{\star},\ell,0\right)Q(\ell,k\Delta t) + \sum_{k=0}^{n}\omega_{n-k}\left(\hat{Q}^{\star},\ell,0\right)w'(\ell,k\Delta t)$$
$$= -\int_{0}^{\ell}\sum_{k=0}^{n}\omega_{n-k}\left(\hat{\varphi}^{\star},x,0\right)q(x,k\Delta t)\,\mathrm{d}x \tag{3.26c}$$

$$\frac{w'(\ell,n\Delta t)}{2} - \sum_{k=0}^{n}\left[\omega_{n-k}\left(\hat{\varphi}'^{\star},x,\ell\right)M(x,k\Delta t) + \omega_{n-k}\left(\frac{\partial\hat{Q}^{\star}}{\partial x},x,\ell\right)w(x,k\Delta t)\right]_{x=0}^{\ell}$$
$$-\sum_{k=0}^{n}\omega_{n-k}\left(\hat{\varphi}^{\star},\ell,0\right)Q(0,k\Delta t) - \sum_{k=0}^{n}\omega_{n-k}\left(\hat{Q}^{\star},\ell,0\right)w'(0,k\Delta t)$$
$$= -\int_{0}^{\ell}\sum_{k=0}^{n}\omega_{n-k}\left(\hat{\varphi}^{\star},x,\ell\right)q(x,k\Delta t)\,\mathrm{d}x \tag{3.26d}$$

for four unknowns are available. Therefore, all boundary data, i.e., the deflection, the slope, the moment, and the shear force at $y=0$ and $y=\ell$, can be determined for every time step $n\Delta t$. But, at inner points, $0<y<\ell$, with equations (3.18) and (3.24) only the deflection and the slope, respectively, is available. Because the points y and x do not coincide for $0<y<\ell$ in equations (3.18) and (3.24) these equations are differentiable at every point $0<y<\ell$. Thus, the moment $M(y)$ is given by the derivative of equation (3.24) with respect to y

$$M(y) = -EI\frac{\mathrm{d}w'}{\mathrm{d}y}$$
$$= \int_{0}^{\ell}\sum_{k=0}^{n}\omega_{n-k}\left(\hat{M}^{\star},x,y\right)q(x,k\Delta t)\,\mathrm{d}x + \sum_{k=0}^{n}\left[\omega_{n-k}\left(\hat{M}^{\star},x,y\right)Q(x,k\Delta t)\right.$$
$$+\omega_{n-k}\left(\frac{\partial\hat{Q}^{\star}}{\partial y},x,y\right)EIw'(x,k\Delta t) + \omega_{n-k}\left(\frac{\partial^2\hat{Q}^{\star}}{\partial y^2},x,y\right)EIw(x,k\Delta t)$$
$$\left.-\,\omega_{n-k}\left(\hat{Q}^{\star},x,y\right)M(x,k\Delta t)\right]_{x=0}^{\ell} \tag{3.27}$$

and a second differentiation leads to the shear force

$$
\begin{aligned}
Q(y) &= \frac{\mathrm{d}M}{\mathrm{d}y} \\
&= \sum_{k=0}^{n} \left[\omega_{n-k}\left(\frac{\partial^2 \hat{Q}^\star}{\partial y^2},x,y\right) EIw'(x,k\Delta t) + \omega_{n-k}\left(\frac{\partial^3 \hat{Q}^\star}{\partial y^3},x,y\right) EIw(x,k\Delta t) \right. \\
&\quad \left. - \omega_{n-k}\left(\hat{Q}^\star,x,y\right) Q(x,k\Delta t) - \omega_{n-k}\left(\frac{\partial \hat{Q}^\star}{\partial y},x,y\right) M(x,k\Delta t) \right]_{x=0}^{\ell} \\
&\quad - \int_0^{\ell} \sum_{k=0}^{n} \omega_{n-k}\left(\hat{Q}^\star,x,y\right) q(x,k\Delta t)\,\mathrm{d}x
\end{aligned} \tag{3.28}
$$

at the inner points $0 < y < \ell$. For the determination of two integration weights in (3.27) and (3.28), the corresponding fundamental solution in Laplace domain are achieved by derivations of the fundamental solution for the shear force (3.10d)

$$
\frac{\partial^2}{\partial y^2}\hat{Q}^\star(x,y) = \frac{\lambda^2}{4}\left[\cosh(\lambda|x-y|) - \cos(\lambda|x-y|)\right](2H(x-y)-1) \tag{3.29a}
$$

$$
\frac{\partial^3}{\partial y^3}\hat{Q}^\star(x,y) = -\frac{\lambda^3}{4}\left[\sinh(\lambda|x-y|) + \sin(\lambda|x-y|)\right] . \tag{3.29b}
$$

Now, the deflection, the slope, the moment, and the shear force at any point y on the beam are available for every time $t = n\Delta t$.

3.2 Numerical example

Two examples will show the accuracy of the proposed method. In the first example, a fixed-simply supported beam, the influence of the time step size and the underlying multistep method is studied. The second example, a viscoelastically supported beam, shows the influence of the viscoelastic support on the wave propagation in the beam.

For all tests, the parameters of the convolution quadrature are chosen as suggested in Sect. 2.2.2: $L = N$ and $\mathscr{R}^N = \sqrt{\varepsilon}$ with $\varepsilon = 10^{-10}$.

3.2.1 Fixed-simply supported beam

Consider a fixed-simply supported beam which is subjected to a suddenly applied point load $q(x,t) = 1\,\mathrm{N/m}\,H(t)\,\delta(x-\ell/2)$ at its midspan. The geometry and the material data are specified in Fig. 3.2. With the proposed method the four unknowns, the deflection, the slope, the moment, and the shear force are calculated. In Fig. 3.3, the deflection w and, in Fig. 3.4, the slope w' for different time step size Δt are depicted versus time t. All values are normalized by the corresponding result of a static calculation, i.e., the deflection by $w_{static}(y = 0.75\ell) = 0.007\,\mathrm{m}$ and the slope by $w'_{static}(y = 0.75\ell) = -0.021$. As expected, due to the permanent loading with

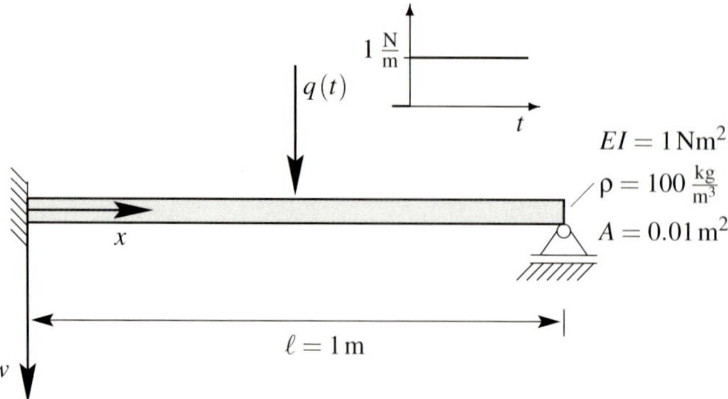

Fig. 3.2. Fixed-simply supported beam

respect to time $q(x,t) = 1\,\mathrm{N/m}\,H(t)\,\delta\left(x - \frac{\ell}{2}\right)$, the results oscillate between 0 and the doubled static values. Three extreme cases choosing the time step size Δt are presented in Fig. 3.3 and Fig. 3.4. The largest value $\Delta t = 0.03\,\mathrm{s}$ is for a proper representation of the results time history not suitable and leads to numerical damping. The decay of the deflection or slope, respectively, has to be numerical damping because no material or geometrical damping is modeled. Obviously, the value $\Delta t = 0.01\,\mathrm{s}$ for the deflection and $\Delta t = 0.005\,\mathrm{s}$ for the slope gives very accurate and stable results, whereas the smallest step size $\Delta t = 0.005\,\mathrm{s}$ or $\Delta t = 0.001\,\mathrm{s}$ tends to overshooting, i.e., the values are smaller than 0 or larger than 2. Especially, the slope in Fig. 3.4 corresponding to the smallest time step size indicates numerical instabilities. Summarizing, the results are satisfactory if the time step size is small enough.

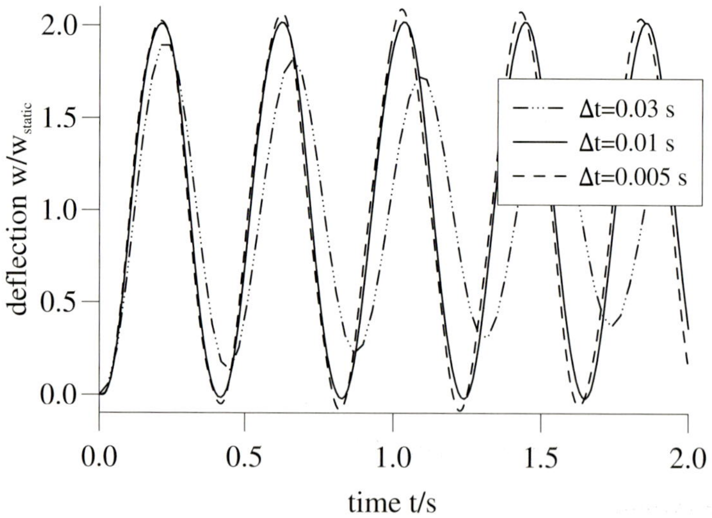

Fig. 3.3. The deflection w at $y = 0.75\ell$ versus time t:Influence of time step size Δt

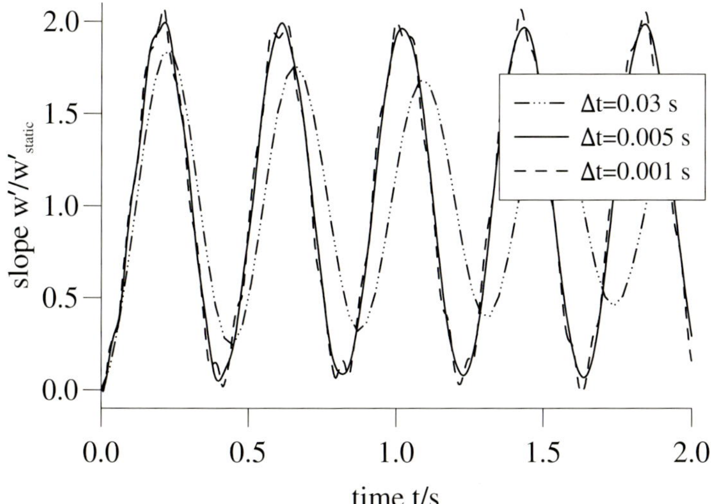

Fig. 3.4. The slope w' at $y = 0.75\ell$ versus time t:Influence of time step size Δt

Next, in Fig. 3.5, the moment and, in Fig. 3.6, the shear force in the fixed support at $y = 0$ is plotted versus time for different time step sizes. The results are

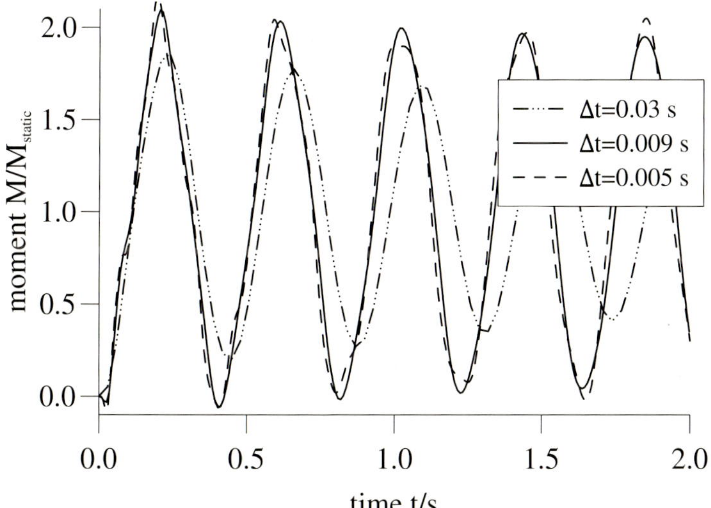

Fig. 3.5. The moment M at $y = 0$ versus time t: Influence of time step size Δt

normalized to the values of a corresponding static calculation, i.e., the moment by $M_{static}\,(y = 0) = -0.1875\,\text{Nm}$ and the shear force by $Q_{static}\,(y = 0) = 0.6875\,\text{N}$.

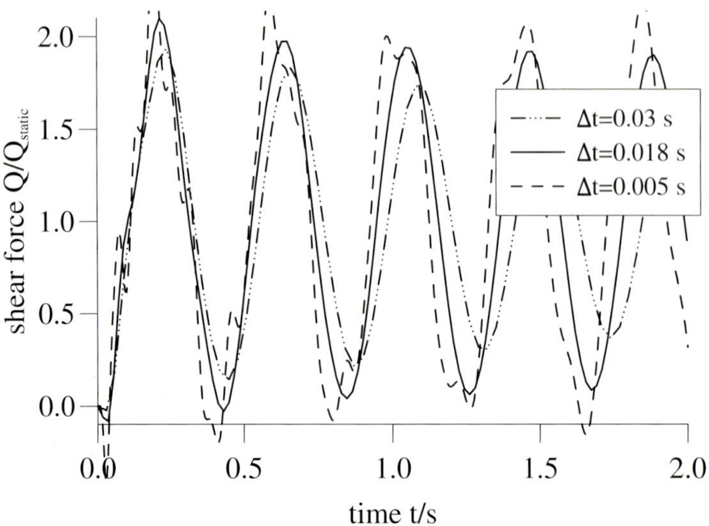

Fig. 3.6. The shear force Q at $y = 0$ versus time t: Influence of time step size Δt

Similar to Figs. 3.3 and 3.4, the results of moment and shear force are presented for three different time step sizes Δt. As before, the large time step size $\Delta t = 0.03$ s is too coarse for an accurate approximation of the time history and, therefore, leads to numerical damping. Contrary to deflection and slope, the moment and the shear force are more sensitive to the time step size. There is no optimal time step size as before characterized by nearly no numerical damping and no overshooting. For the moment and shear force only a "best compromise" between numerical damping and no negative values can be found: this is $\Delta t = 0.009$ s for the moment and $\Delta t = 0.018$ s for the shear force.

Values for the moment or for the shear force outside the interval $[0, 2]$ point to instability because there is no physical reason for these values. They can only be caused by numerics. The shear force results (Fig. 3.6) are worse than the results for the moment (Fig. 3.5), but this can not be generalized. At different locations y or for different boundary conditions this can be different. However, for all cases the deflection and the slope are approximated better than the moment and the shear force. This could probably be improved using even for the determination of the boundary data two additional integral equations for the moment and the shear force.

The next interesting question concerns the influence of the underlying multistep method. As mentioned in Chap. 2.1, the convolution quadrature method requires an $A(\alpha)$-stable multistep method which is stable at infinity, i.e., a disturbance of the solution is damped out by the method. Here, the BDF 1, BDF 2, BDF 3, and the trapezoidal rule are applied. The BDF 1, BDF 2, and the trapezoidal rule are even A-stable, whereas the BDF 3 is only $A(\alpha)$-stable. Further, all three BDF's fulfill the stability at infinity contrary to the trapezoidal rule. In Fig. 3.7, now, the deflection $w(y = 0.75\ell)$ is depicted versus time for the four mentioned multistep methods.

For every method an optimal time step size is used. In case of BDF 3 no stable

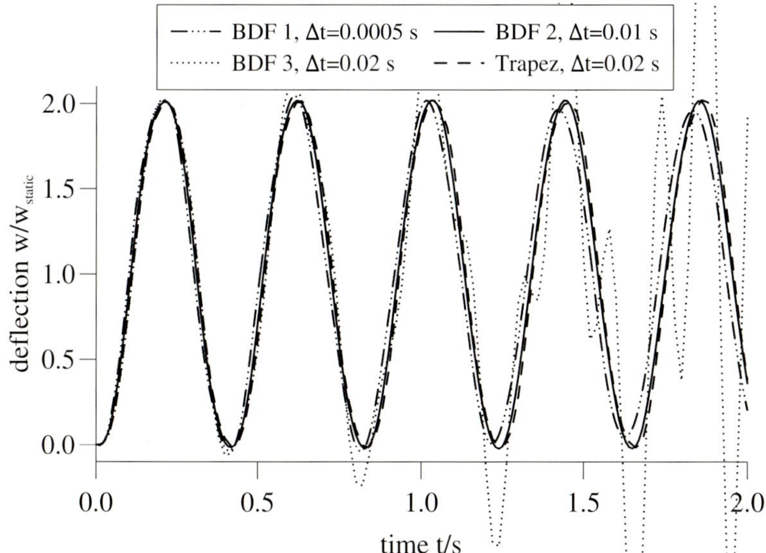

Fig. 3.7. The deflection w at $y = 0.75\ell$ versus time t: Influence of the underlying multistep method

solution is found, and, therefore, the "best" solution is plotted. Clearly, this method is not suitable for the beam integral equation. However, the three other methods give comparable results with a tendency to the BDF 2 and the trapezoidal rule. For the BDF 1 a very small Δt has to be chosen to minimize numerical damping. Summarizing, for this application an A-stable method is necessary but it can be waived the stability at infinity.

3.2.2 Fixed-free viscoelastic supported beam

Consider a fixed-free supported beam resting on a viscoelastic layer which is subjected to a suddenly applied point load $q(x,t)$ at its end. The geometry and the material data are specified in Fig. 3.8. To study wave propagation phenomenon in the beam a impulse load modeled by the "hat"-function with respect to time $q(x,t) = 1\,\mathrm{N/m}\ (H(t) - H(t-0.2\,\mathrm{s}))\,\delta(x-\ell)$ is used. Because the constitutive equation for the beam is Hook's law the oscillation of the deflection around the origin is damped only by the viscous foundation. This effect can be observed in Fig. 3.9. There, the case of no foundation denoted by *no support* is compared with an elastic, a viscous, and a viscoelastic support. For the elastic support no damping is involved only a stiffening of the system due to the springs is expected and Fig. 3.9 confirms this by smaller amplitudes of the deflection and faster wave velocity. The

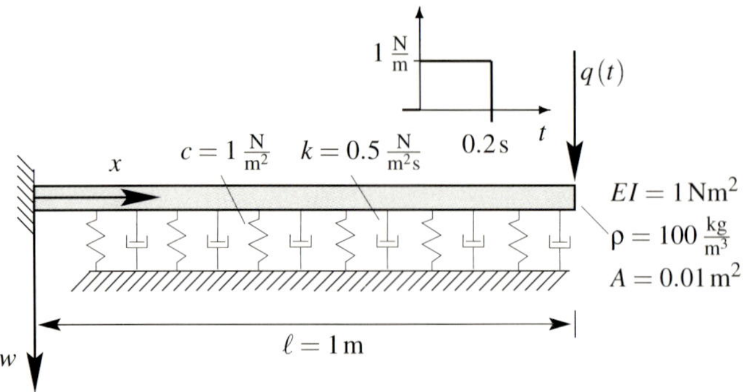

Fig. 3.8. Fixed-free viscoelastically supported beam

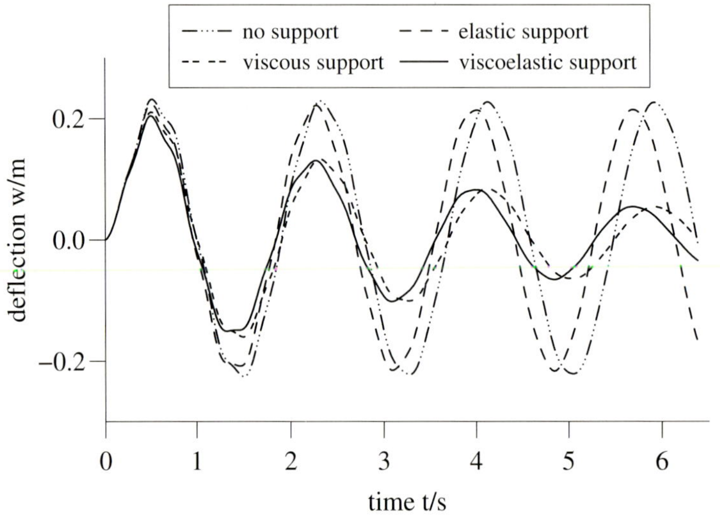

Fig. 3.9. The deflection w at $y = 0.75\ell$ versus time t: Different models of the support

purely viscous support dissipates the energy in the oscillating beam, i.e., the deflection decreases to zero. The combination of both effects is the viscoelastic support leading to damped oscillations with smaller absolute values.

Both examples had shown that with the proposed integral equation method a reliable formulation for the wave propagation in beams has been developed. Remarkable is that the solution with respect to the spatial variable is exact contrary to the time behavior which is approximated by the convolution quadrature method.

4. Time domain boundary element formulation

In Chap. 3 the wave propagation in a beam, a one-dimensional continuum, was treated. Now, in this chapter the integral equation and, finally, a boundary element formulation for a two- (2-d) or three-dimensional (3-d) continuum will be deduced.

For this, first, the problem has to be defined. Consider a domain Ω with boundary $\Gamma = \Gamma_t + \Gamma_u$ loaded with a body force $\mathbf{b}$. On Γ_t the traction calculated by Cauchy's theorem

$$\mathbf{t} = \sigma\mathbf{n} \qquad t_i = \sigma_{ij} n_j \tag{4.1}$$

and on the remaining part of the boundary Γ_u the displacement $\mathbf{u}$ is prescribed (see

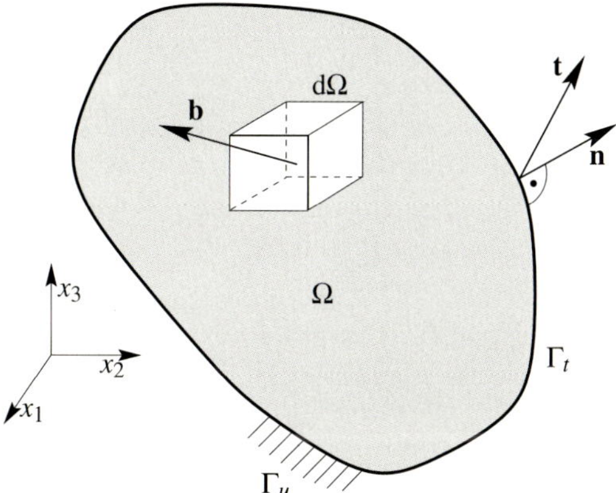

Fig. 4.1. General configuration of the problem

Fig. 4.1). The outward normal is denoted by $\mathbf{n}$ and a Cartesian coordinate system is used. In the following, small displacement gradients are assumed allowing not to distinguish between Lagrangian and Eulerian coordinate system, i.e., a linear strain-displacement relation

$$\varepsilon_{ij} = \frac{1}{2}\left(u_{i,j} + u_{j,i}\right) \tag{4.2}$$

is assumed with ε_{ij} denoting the components of the strain tensor. The dynamic equilibrium is

$$\sigma_{ij,j} + b_i = \rho \ddot{u}_i \tag{4.3}$$

with the Cauchy stress tensor σ_{ij} and the mass density ρ. The double dot on the displacement denotes the acceleration.

To complete the problem description, the connection between the stress and strain a constitutive equation must be formulated. In this chapter, an elastic homogeneous material behavior will be assumed

$$\sigma_{ij} = C_{ijkl}\varepsilon_{kl} \ , \tag{4.4}$$

with the fourth order material tensor C_{ijkl}. This material tensor has the symmetries

$$C_{ijkl} = C_{jikl} = C_{ijlk} \tag{4.5}$$

with respect to the first and second couple of indices. Due to these symmetries and with the linear geometry relation (4.2) the stress can be expressed

$$\sigma_{ij} = C_{ijkl}\frac{1}{2}\left(u_{k,l} + u_{l,k}\right) = \frac{1}{2}\left(C_{ijkl}u_{k,l} + C_{ijlk}u_{l,k}\right) = C_{ijkl}u_{k,l} \ . \tag{4.6}$$

Next, the integral formulation of the above given problem can be derived.

4.1 Integral equation for elastodynamics

The mostly used possibilities to get the integral equation are either a formal mathematical way using weighted residuals or a more physical using the reciprocal work theorem. To present both possibilities, in this chapter the weighted residual statement will be used, whereas later for a viscoelastic constitutive equation the way using the reciprocal work theorem will be demonstrated.

The weighted residual statement for elastodynamics is achieved by equating the inner product of the dynamic equilibrium (4.3) and the fundamental solution for the displacements U_{ij} in the spatial variable and the convolution with respect to the time variable to zero, i.e.,

$$\int_0^t \int_\Omega \left(\sigma_{ik,k}(\tau,\mathbf{x}) + b_i(\tau,\mathbf{x}) - \rho\ddot{u}_i(\tau,\mathbf{x})\right) U_{ij}(t-\tau,\mathbf{x},\mathbf{y})\,\mathrm{d}\Omega\mathrm{d}\tau = 0 \ . \tag{4.7}$$

This essentially forces the error associated with the satisfaction of the governing differential equation (4.3) to be orthogonal to U_{ij}. Physically interpreted this corresponds that the error introduced solving the differential equation (4.3) is zero averaged over the domain Ω and time t. To choose the fundamental solution U_{ij} as the weighting function is at this point arbitrary, indeed, every smooth function could be used. But, to arrive finally at the boundary element formulation the weighting function must be the fundamental solution.

The constitutive equation (4.4) is valid for an anisotropic homogeneous elastic media, however, due to the lack of fundamental solutions for the general anisotropic case, the following is restricted to an isotropic homogeneous continuum. For the sake of brevity, in the following, the argument list will be skipped and the convolution will be denoted by $*$.

Next, the three parts in the brackets are treated individually. Partial integration of the first term using the divergence theorem and the Cauchy theorem (4.1) yields

$$\begin{aligned}\int_\Omega \sigma_{ik,k} * U_{ij}\mathrm{d}\Omega &= \int_\Gamma \sigma_{ik}n_k * U_{ij}\mathrm{d}\Gamma - \int_\Omega \sigma_{ik} * U_{ij,k}\mathrm{d}\Omega \\ &= \int_\Gamma t_i * U_{ij}\mathrm{d}\Gamma - \int_\Omega \sigma_{ik} * U_{ij,k}\mathrm{d}\Omega\,.\end{aligned} \tag{4.8}$$

A second partial integration is applied to the last integral in (4.8)

$$\begin{aligned}\int_\Omega \sigma_{ik} * U_{ij,k}\mathrm{d}\Omega &= \int_\Omega C_{ikmn}u_{m,n} * U_{ij,k}\mathrm{d}\Omega \\ &= \int_\Gamma u_m * C_{ikmn}U_{ij,k}n_n\mathrm{d}\Gamma - \int_\Omega u_m * C_{ikmn}U_{ij,kn}\mathrm{d}\Omega \\ &= \int_\Gamma u_i * T_{ij}\mathrm{d}\Gamma - \int_\Omega u_i * \Sigma_{ijk,k}\mathrm{d}\Omega\end{aligned} \tag{4.9}$$

using the property of the material tensor (4.6) and Σ_{ijk} denotes the fundamental solution for the stress tensor calculated by the constitutive equation (4.6) with the fundamental solutions for the displacements. These manipulations shift the divergence operator from the stress tensor σ to the fundamental solution.

The inertia term in the integral (4.7) is also mapped on the fundamental solution U_{ij} using two partial integrations with respect to time

$$\begin{aligned}\ddot{u}_i * U_{ij} &= \int_0^t \frac{\partial^2}{\partial\tau^2}u_i(\tau)\,U_{ij}(t-\tau)\,\mathrm{d}\tau \\ &= \left[\frac{\partial}{\partial\tau}u_i(\tau)\,U_{ij}(t-\tau) - u_i(\tau)\,\frac{\partial}{\partial\tau}U_{ij}(t-\tau)\right]_0^t + \int_0^t u_i(\tau)\,\frac{\partial^2}{\partial\tau^2}U_{ij}(t-\tau)\,\mathrm{d}\tau\,.\end{aligned} \tag{4.10}$$

With the assumption of vanishing initial conditions for the displacements and the fundamental solutions

$$u_i(t=0,\mathbf{x}) = \dot{u}_i(t=0,\mathbf{x}) = U_{ij}(t-\tau=0,\mathbf{x},\mathbf{y}) = \dot{U}_{ij}(t-\tau=0,\mathbf{x},\mathbf{y}) = 0 \tag{4.11}$$

the integral (4.7) is finally

$$0 = \int_\Omega b_i * U_{ij}\mathrm{d}\Omega + \int_\Omega u_i * \left(\Sigma_{ijk,k} - \rho\ddot{U}_{ij}\right)\mathrm{d}\Omega + \int_\Gamma t_i * U_{ij}\mathrm{d}\Gamma - \int_\Gamma u_i * T_{ij}\mathrm{d}\Gamma\,. \tag{4.12}$$

Now, inserting the condition for determining a fundamental solution (see appendix B.1)

$$\Sigma_{ijk,k}(t-\tau,\mathbf{x},\mathbf{y}) - \rho\ddot{U}_{ij}(t-\tau,\mathbf{x},\mathbf{y}) = -\delta(t-\tau)(\mathbf{x}-\mathbf{y}) \tag{4.13}$$

and using the properties of the Dirac distribution δ the integral equation is achieved

$$\begin{aligned} u_j(t,\mathbf{y}) = &\int_0^t \int_\Gamma \left[t_i(\tau,\mathbf{x})U_{ij}(t-\tau,\mathbf{x},\mathbf{y}) - u_i(\tau,\mathbf{x})\,T_{ij}(t-\tau,\mathbf{x},\mathbf{y})\right]\mathrm{d}\Gamma\mathrm{d}\tau \\ &+ \int_0^t \int_\Omega b_i(\tau,\mathbf{x})\,U_{ij}(t-\tau,\mathbf{x},\mathbf{y})\,\mathrm{d}\Omega\mathrm{d}\tau \qquad \mathbf{y}\in\Omega \end{aligned} \tag{4.14}$$

with the boundary data displacement u_i and traction t_i as the only unknowns.

At every point $\mathbf{x} \in \Gamma$ either displacements or tractions must be prescribed for a well posed problem. To determine the missing boundary data, respectively, $\mathbf{y}$ has to be moved on the boundary Γ. Due to the singular behavior of the fundamental solutions if $\mathbf{y}$ approach $\mathbf{x}$ this is only possible in a limiting procedure as extensively described in the literature (see, e.g., [41, 76]). Hence, here, this limiting process is only briefly recalled.

Necessary for this limit is the knowledge about the singular behavior of the fundamental solutions. Inserting the series representation of the exponential function $\mathrm{e}^{-rs/c_i} = \sum_{n=0}^{\infty}(-rs/c_i)^n\,1/n!$ with $r_i = x_i - y_i$ and $r = \sqrt{r_i r_i}$ in the Laplace transformed fundamental solutions (B.1) and (B.2) and a subsequent rearrangement following powers of r yields for the displacement solution

$$\hat{U}_{ij}(\mathbf{x},\mathbf{y},s) = \underbrace{\frac{1+\nu}{8\pi E(1-\nu)}\left\{r_{,i}r_{,j} + \delta_{ij}(3-4\nu)\right\}\frac{1}{r}}_{\text{static fundamental solution}} + \mathcal{O}\left(r^0\right) \tag{4.15}$$

and for the traction solution

$$\begin{aligned} \hat{T}_{ij}(\mathbf{x},\mathbf{y},s) = &\underbrace{\frac{-1}{8\pi(1-\nu)}\left\{\left[(1-2\nu)\delta_{ij} + 3r_{,i}r_{,j}\right]r_{,n} - (1-2\nu)\left(r_{,j}n_i - r_{,i}n_j\right)\right\}\frac{1}{r^2}}_{\text{static fundamental solution}} \\ &+ \mathcal{O}\left(r^0\right), \end{aligned} \tag{4.16}$$

with Young's modulus E and Poisson's ratio ν. The singular part of the functions (4.15) and (4.16) is identified as the elastostatic fundamental solution, respectively. Therefore, the displacement fundamental solution is weakly and the traction fundamental solution strongly singular. Though the series expansion in (4.15) and (4.16) is presented in Laplace domain and, therefore, the singular behavior is only known in the transformed domain, this result can be transfered to the time domain, because the singular parts are independent of the Laplace variable s and consequently independent of time. Thus, also, the time-dependent fundamental solutions behave like the elastostatic ones in the limit $\mathbf{y} \to \mathbf{x}$. A proof of these singularities direct in time domain can be found in [37]. The same splitting of the fundamental solutions in a singular elastostatic part and a regular "dynamic" part is also possible for the 2-d solutions.

Now, to perform the limiting process the boundary is deformed as shown in Fig. 4.2 for the 2-d case. In 3-d the respective spherical deformation of the boundary has to be used. With this deformation the point $\mathbf{y}$ resides inside the domain Ω. If ε tends to zero the point $\mathbf{y}$ moves on the boundary. Therefore, the integral over Γ in equation (4.14) is divided in two parts as depicted in Fig. 4.2. The first part on the right hand side of this equation can either be regularized by a coordinate transformation for $t_i(t,\mathbf{x}) * U_{ij}(t,\mathbf{x},\mathbf{y})$ and has to be defined in the sense of a Cauchy Principal Value (CPV) for $u_i(t,\mathbf{x}) * T_{ij}(t,\mathbf{x},\mathbf{y})$ due to the weak and strong singularity, respectively. In the second integral only the term $\lim\limits_{\varepsilon\to 0} \int_{\Gamma_\varepsilon} u(t,\mathbf{x}) * T_{ij}(t,\mathbf{x},\mathbf{y})\, \mathrm{d}\Gamma_\varepsilon$ has to be discussed, because the first part of this integral is zero due to the weak singularity. Together with the integral free displacement term in equation (4.14) the remaining limit yields the integral free term $c_{ij}(\mathbf{y}) = 1 + \lim\limits_{\varepsilon\to 0} \int_{\Gamma_\varepsilon} T_{ij}^{static}(\mathbf{x},\mathbf{y})\, \mathrm{d}\Gamma_\varepsilon$ under the assumption of a Hölder continuous displacement $u_i(t,\mathbf{x})$ [41]. To calculate the integral free term only the elastostatic fundamental solution $T_{ij}^{static}(\mathbf{x},\mathbf{y})$ is necessary, because only this part of $T_{ij}(t,\mathbf{x},\mathbf{y})$ is singular. Therefore, this term is only dependent on the geometry and Poisson's ratio, e.g., for a smooth surface $c_{ij} = 1/2\, \delta_{ij}$. For arbitrary boundary points, e.g., corners, the term c_{ij} is determined following a procedure given in [129]. Finally, these considerations result in the boundary integral equation

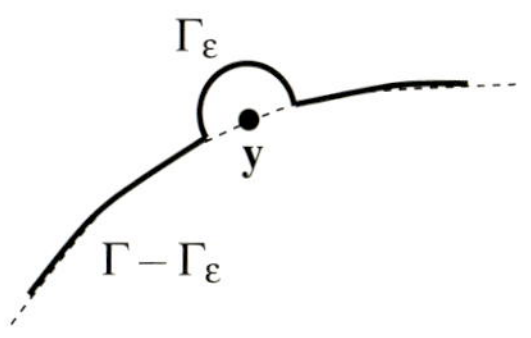

$$\int\limits_{\Gamma} \left[t_i(t,\mathbf{x}) * U_{ij}(t,\mathbf{x},\mathbf{y}) - u_i(t,\mathbf{x}) * T_{ij}(t,\mathbf{x},\mathbf{y})\right] \mathrm{d}\Gamma =$$
$$\lim_{\varepsilon\to 0} \int\limits_{\Gamma-\Gamma_\varepsilon} \left[t_i(t,\mathbf{x}) * U_{ij}(t,\mathbf{x},\mathbf{y}) - u_i(t,\mathbf{x}) * T_{ij}(t,\mathbf{x},\mathbf{y})\right] \mathrm{d}\Gamma$$
$$+ \lim_{\varepsilon\to 0} \int\limits_{\Gamma_\varepsilon} \left[t_i(t,\mathbf{x}) * U_{ij}(t,\mathbf{x},\mathbf{y}) - u_i(t,\mathbf{x}) * T_{ij}(t,\mathbf{x},\mathbf{y})\right] \mathrm{d}\Gamma_\varepsilon$$

Fig. 4.2. Deformed boundary (2-d)

$$\int_0^t \int_\Gamma t_i(\tau,\mathbf{x})\, U_{ij}(t-\tau,\mathbf{x},\mathbf{y})\, \mathrm{d}\Gamma \mathrm{d}\tau - \int_0^t \oint_\Gamma u_i(\tau,\mathbf{x})\, T_{ij}(t-\tau,\mathbf{x},\mathbf{y})\, \mathrm{d}\Gamma \mathrm{d}\tau$$
$$+ \int_0^t \int_\Omega b_i(\tau,\mathbf{x})\, U_{ij}(t-\tau,\mathbf{x},\mathbf{y})\, \mathrm{d}\Omega \mathrm{d}\tau = c_{ij}(\mathbf{y})\, u_i(t,\mathbf{y}) \quad \mathbf{y} \in \Gamma\,. \tag{4.17}$$

With this boundary integral equation – the unknowns are only in the boundary integrals – the displacements on the boundary Γ and subsequent with (4.14) in the domain Ω are given. For an arbitrary domain the integral equation can not be solved analytically and, therefore, a discretization is introduced leading to the boundary element formulation.

4.2 Boundary element formulation for elastodynamics

According to the boundary element method, the boundary surface Γ is discretized by E elements where polynomial spatial shape functions $N_e^f(\mathbf{x})$ with F nodes are defined (see, e.g., [140]). Hence, with the time-dependent nodal values $u_i^{ef}(t)$ and $t_i^{ef}(t)$ the displacements and tractions are approximated, respectively, by

$$u_i(t,\mathbf{x}) = \sum_{e=1}^{E} \sum_{f=1}^{F} N_e^f(\mathbf{x})\; u_i^{ef}(t)\,, \qquad t_i(t,\mathbf{x}) = \sum_{e=1}^{E} \sum_{f=1}^{F} N_e^f(\mathbf{x})\; t_i^{ef}(t)\,. \tag{4.18}$$

Inserting these ansatz functions in the integral equation (4.17) yields

$$\sum_{e=1}^{E} \sum_{f=1}^{F} \left\{ \int_\Gamma U_{ij}(\mathbf{x},\mathbf{y},t)\, N_e^f(\mathbf{x})\, \mathrm{d}\Gamma * t_i^{ef}(t) - \oint_\Gamma T_{ij}(\mathbf{x},\mathbf{y},t)\, N_e^f(\mathbf{x})\, \mathrm{d}\Gamma * u_i^{ef}(t) \right\}$$
$$+ \int_\Omega b_i(t,\mathbf{x}) * U_{ij}(t,\mathbf{x},\mathbf{y})\, \mathrm{d}\Omega = c_{ij}(\mathbf{y})\, u_i(\mathbf{y},t)\,. \tag{4.19}$$

In the domain integral of equation (4.19) no ansatz functions are necessary because the integrand is known and needs only to be integrated numerically over the domain. In some special cases, e.g., for the gravity force, the domain integral can be transformed to the boundary [7].

The next step is the time discretization. Usually ansatz functions also for the time variable are introduced and subsequent the time integration in each time step is performed analytical. This method was first introduced by Mansur [126] and is briefly sketched in appendix B.2. In the following, this method will be called *classical*. Here, a different time stepping procedure is introduced based on the convolution quadrature method.

When the time period t is discretized by N time steps of equal duration Δt, the convolution integral between the fundamental solutions $U_{ij}(\mathbf{x},\mathbf{y},t)$ or $T_{ij}(\mathbf{x},\mathbf{y},t)$ and the nodal values $t_i^{ef}(t)$ or $u_i^{ef}(t)$, respectively, is approximated by the convolution quadrature formula (2.18). This results in the following boundary element time-stepping formulation for $n = 0, 1, \ldots, N$

$$\sum_{e=1}^{E}\sum_{f=1}^{F}\sum_{k=0}^{n}\left\{\omega_{n-k}^{ef}\left(\hat{U}_{ij},\mathbf{y},\Delta t\right)t_i^{ef}(k\Delta t) - \omega_{n-k}^{ef}\left(\hat{T}_{ij},\mathbf{y},\Delta t\right)u_i^{ef}(k\Delta t)\right\} + \sum_{k=0}^{n}\int_{\Omega}\omega_{n-k}^{\Omega}\left(\hat{U}_{ij},\mathbf{y},\Delta t\right)b_i(k\Delta t,\mathbf{x})\,\mathrm{d}\Omega = c_{ij}(\mathbf{y})\,u_j(\mathbf{y},n\Delta t)\,. \tag{4.20}$$

with the integration weights corresponding to equation (2.22)

$$\omega_n^{ef}\left(\hat{U}_{ij},\mathbf{y},\Delta t\right) = \frac{\mathscr{R}^{-n}}{L}\sum_{\ell=0}^{L-1}\int_{\Gamma}\hat{U}_{ij}\left(\mathbf{x},\mathbf{y},\frac{\gamma\left(\mathscr{R}\mathrm{e}^{-\mathrm{i}\ell\frac{2\pi}{L}}\right)}{\Delta t}\right)N_e^f(\mathbf{x})\,\mathrm{d}\Gamma\,\mathrm{e}^{-\mathrm{i}n\ell\frac{2\pi}{L}} \tag{4.21a}$$

$$\omega_n^{ef}\left(\hat{T}_{ij},\mathbf{y},\Delta t\right) = \frac{\mathscr{R}^{-n}}{L}\sum_{\ell=0}^{L-1}\oint_{\Gamma}\hat{T}_{ij}\left(\mathbf{x},\mathbf{y},\frac{\gamma\left(\mathscr{R}\mathrm{e}^{-\mathrm{i}\ell\frac{2\pi}{L}}\right)}{\Delta t}\right)N_e^f(\mathbf{x})\,\mathrm{d}\Gamma\,\mathrm{e}^{-\mathrm{i}n\ell\frac{2\pi}{L}} \tag{4.21b}$$

$$\omega_n^{\Omega}\left(\hat{U}_{ij},\mathbf{y},\Delta t\right) = \frac{\mathscr{R}^{-n}}{L}\sum_{\ell=0}^{L-1}\hat{U}_{ij}\left(\mathbf{x},\mathbf{y},\frac{\gamma\left(\mathscr{R}\mathrm{e}^{-\mathrm{i}\ell\frac{2\pi}{L}}\right)}{\Delta t}\right)\mathrm{e}^{-\mathrm{i}n\ell\frac{2\pi}{L}}\,. \tag{4.21c}$$

Note, contrary to the known formulations in time domain, here, the calculation of the integration weights (4.21) is only based on the Laplace transformed fundamental solution, i.e., this method is also applicable to visco- or poroelasticity where the fundamental solutions in time domain are not available.

Further, the spatial integrations in this formulation are performed over smooth functions – the fundamental solutions in Laplace domain are composed of exponential functions. Contrary, in the classical formulation not continuous functions has to be integrated, i.e., the spatial integration has to be performed over the wave fronts. Here, the spatial integration in the weights (4.21) over each boundary element Γ is realized by Gaussian quadrature. Only when $\mathbf{x}$ approaches $\mathbf{y}$, the respective integral of (4.21a) is regularized with a coordinate transformation and the integral in (4.21b) for the tractions (CPV) is evaluated following the procedure proposed by Guiggiani and Gigante [101]. In order to arrive in equation (4.20) at systems of algebraic equations, collocation is used at every node of the shape functions $N_e^f(\mathbf{x})$.

The integration weights ω_{n-k}^{ef} in equation (4.20) are only dependent on the difference $n-k$ not on n, i.e., the integration weights are calculated only for the relative time $(n-k)\Delta t = t - \tau$ and not for the absolute time $n\Delta t = t$. This property is analogous to the classical elastodynamic time domain boundary element formulation (see, e.g., [76] or appendix B.2) and can be used to establish a recursion formula for $n = 1, 2, \ldots, N$ $(m = n-k)$

$$\omega_0(\mathbf{C})\,\mathbf{d}^n = \omega_0(\mathbf{D})\,\bar{\mathbf{d}}^n + \sum_{m=1}^{n}\left(\omega_m(\mathbf{U})\,\mathbf{t}^{n-m} - \omega_m(\mathbf{T})\,\mathbf{u}^{n-m}\right) + \omega_n(\mathbf{b}) \tag{4.22}$$

analogous to equation (B.16) of the classical formulation. In the above formula, the time-dependent integration weights ω_m contain the Laplace transformed fundamental solutions of the displacements $\mathbf{U}$ (4.21a) and of the tractions $\mathbf{T}$ (4.21b), respectively. Similarly, $\omega_0(\mathbf{C})$ and $\omega_0(\mathbf{D})$ are the corresponding integration weights of the first time step related to the unknown boundary data $\mathbf{d}^n$ and the known boundary data $\bar{\mathbf{d}}^n$ in time step n, respectively. The vector $\omega_n(\mathbf{b})$ denotes the integrated volume forces (see equation (4.21c)). In representation (4.22) of the integral equation (4.20), it is observed that only N matrices $\omega_m(\mathbf{T}), \omega_m(\mathbf{U})$ and, therefore, only N integration weights have to be determined. Finally, a direct equation solver is applied to equation (4.22), where a LU decomposition is preferable to do only once the decomposition for all time steps $n = 0, 1, \ldots, N$.

4.3 Validation of proposed method: Wave propagation in a rod

In order to validate the proposed time stepping BE formulation wave propagation in a rod is studied with respect to the influence of

- the spatial discretization,
- the time step size and
- the underlying multistep method.

For all tests, the parameters of the convolution quadrature are chosen as suggested by Lubich [119]: $L = N$ and $\mathcal{R}^N = \sqrt{\varepsilon}$ with $\varepsilon = 10^{-10}$. This choice was confirmed by the numerical studies with the test functions (2.23) and (2.24) in Sect. 2.2. Because the fundamental solutions in elastodynamics (B.1) and (B.2) are superpositions of (2.23) and (2.24) all results concerning the test functions (2.23) and (2.24) can be assigned to the fundamental solutions itself [161, 155]. Due to this, it is assumed that also the proposed BE formulation behaves like the test functions (2.23) and (2.24) and, therefore, the parameters determined in Sect. 2.2 are a good choice for the BE formulation as well. The following study will confirm this.

The problem geometry and the material data of the 3-d rod are shown in Fig. 4.3. The rod is fixed on one end, and excited by a pressure jump according to a unit step function $t_y(\mathbf{x},t) = 1\,\mathrm{N/m^2}\,H(t)$ on the other free end. The remaining surfaces are traction free. The material data represent steel with the exception of Poisson's ratio which is chosen $\nu = 0$. This artificial value is taken in order to model a one-dimensional behavior to compare the results with the 1-d analytical solution of longitudinal waves. This solution is found in [99] for the displacement

$$u(y,t) = \frac{t_y}{\rho c}\sum_{n=0}^{\infty}(-1)^n\left[\left(t - \frac{(2n+1)\ell - y}{c}\right) H\left(t - \frac{(2n+1)\ell - y}{c}\right) - \left(t - \frac{(2n+1)\ell + y}{c}\right) H\left(t - \frac{(2n+1)\ell + y}{c}\right)\right] \tag{4.23}$$

and the normal stress

$$\sigma(y,t) = t_y \sum_{n=0}^{\infty} (-1)^n \left[H\left(t - \frac{(2n+1)\ell - y}{c}\right) - H\left(t - \frac{(2n+1)\ell + y}{c}\right) \right] \tag{4.24}$$

with ℓ denoting the length of the rod and the 1-d wave velocity is $c = \sqrt{E/\rho}$.

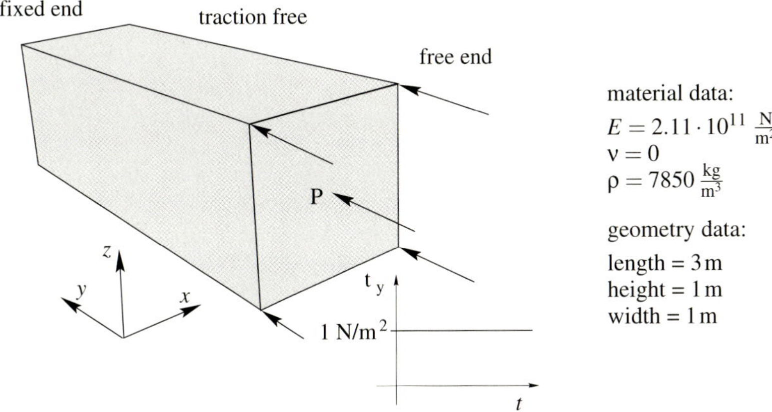

Fig. 4.3. Step function excitation of a free–fixed steel rod

4.3.1 Influence of the spatial and time discretization

First, the results for three discretizations are compared to study the influence of the spatial discretization. The three chosen discretizations are shown in Fig. 4.4, a very coarse mesh with 56 triangles on 30 nodes (mesh 1), a finer mesh with 112 triangles on 58 nodes (mesh 2), and a non uniform mesh with refinement towards the edges with 324 triangles on 164 nodes. For all meshes linear spatial shape functions on triangles are used. In the following, all results are normalized by their corresponding static values, i.e., the displacements by $u_{Static} = 1.4218 \cdot 10^{-11}\,\mathrm{m}$ and the tractions by $t_{Static} = 1\,\mathrm{N}/\mathrm{m}^2$, respectively.

For all three meshes the displacement at the midpoint of the free end (point P) and the traction at the midpoint of the fixed end in longitudinal direction is plotted versus time in Figs. 4.5 and 4.6, respectively. The results for the displacement as well as for the traction are in good agreement with the 1-d solution, whereas the coarse mesh leads to the largest errors as expected. Mesh 3 reproduces the jumps in the traction solution best. However, for the displacement mesh 2 or even mesh 1 is sufficient.

For all calculations a BDF 2 as the underlying multistep method and the optimal time step size Δt corresponding to every mesh is used. Optimal in this context means

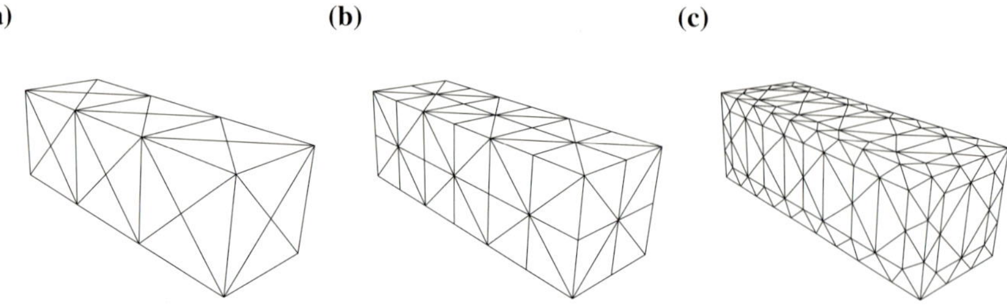

Fig. 4.4. Different discretizations of the 3-d rod **(a)** mesh 1 **(b)** mesh 2 **(c)** mesh 3

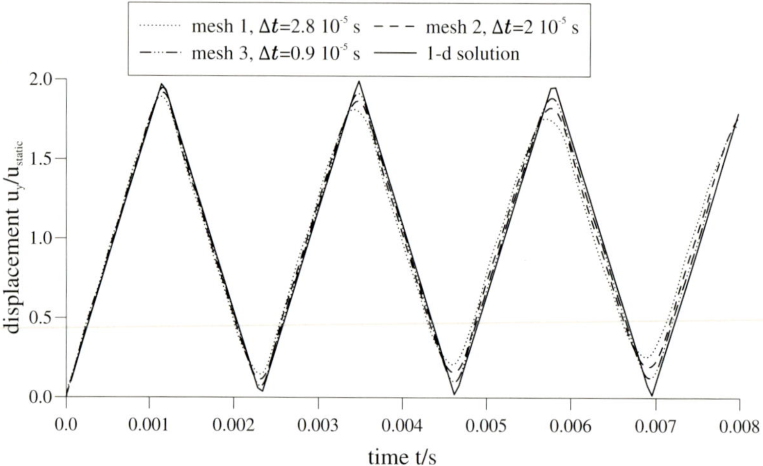

Fig. 4.5. Longitudinal displacement at the free end of the rod versus time: Influence of mesh size

that the results are stable and there is as less as possible numerical damping. This optimal choice depends basically on the wave velocities, i.e., on the material and on the spatial discretization. Therefore, the dimensionless value

$$\beta = \frac{c_1 \Delta t}{r_e} \tag{4.25}$$

is introduced with the characteristic element length r_e. To determine this characteristic element length is in 2-d simply the mean value of all element lengths. But, in 3-d every element has two dimensions and, therefore, it is not clear which length should be chosen. In the following, the mean value of the cathetus of the triangles is chosen, i.e., for mesh 1 $r_e = 1/\sqrt{2}$, for mesh 2 $r_e = 1/2$, and for mesh 3 $r_e = 1/3$.

In most boundary element formulations in time domain, the value β is restricted to a very small range where stable and satisfactory results are achieved. In the classical time domain formulation $0.7 < \beta < 1$ for 3-d is used. The next test will give an answer about this topic for the proposed formulation.

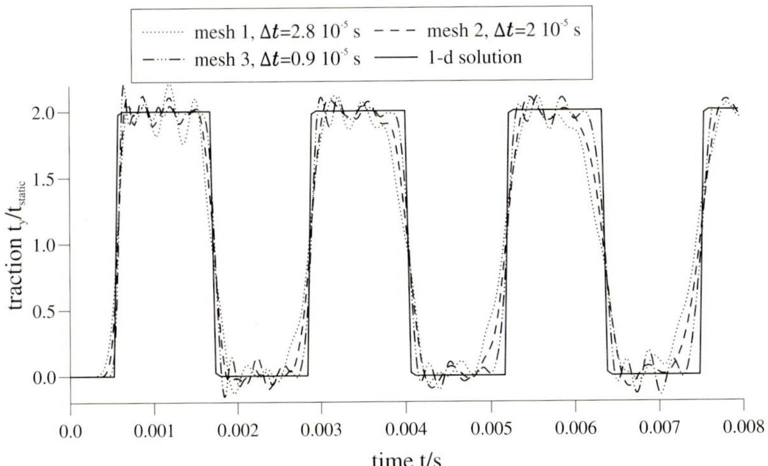

Fig. 4.6. Traction at the fixed end of the rod versus time: Influence of mesh size

The displacements and tractions for different β are depicted in Fig. 4.7 and Fig. 4.8, respectively, using the BDF 2. Clearly indicated, all meshes show a critical time step size, corresponding to $\beta \lesssim 0.15$ for mesh 1 and 2 and $\beta \lesssim 0.07$ for mesh 3, below which the results are unstable. The unstable results are truncated when the solution oscillates too much. The instabilities occur in the traction solution for slightly larger β, respective for larger time steps Δt, than in the displacement solution. But more important, for finer meshes this critical time step size tends to smaller values. Furthermore, the results of mesh 1 strongly and of mesh 2 weakly depend on β, i.e., the solutions deviate more from the 1-d solution if β increases. Contrary, the results using mesh 3 are for all $\beta < 1$ close to the 1-d solution. The stronger dependency on β, respective from the time step size Δt, of mesh 1 or 2 has two reasons. First, the spatial integration on mesh 3 has a much better quality as on mesh 1 or 2 due to the smaller elements. Second, the same β for each mesh represents smaller time step sizes for finer meshes. Because a finer time discretization will always lead to a better approximation of the time history, the time dependency of the displacements and tractions must be better approximated by finer meshes. The results for mesh 3, where nearly no dependence on β is observed, gives reason to the conclusion, that the results of a fine enough mesh are not dependent on the time step size, if $\beta < 1$ is regarded. This limit means, physically interpreted, that the compression wave travels not completely over one element length in one time step.

Not only the time step size also the underlying multistep method influences the results. Therefore, in Fig. 4.9, the longitudinal displacement and, in Fig. 4.10, the traction versus time for different multistep methods using mesh 2 are presented. The linear multistep methods Backward Differential Formulation of first, second, and third order (BDF 1, BDF 2, and BDF 3) are compared with the trapezoidal rule. Figures 4.9 and 4.10 show the results using the relevant optimal time step

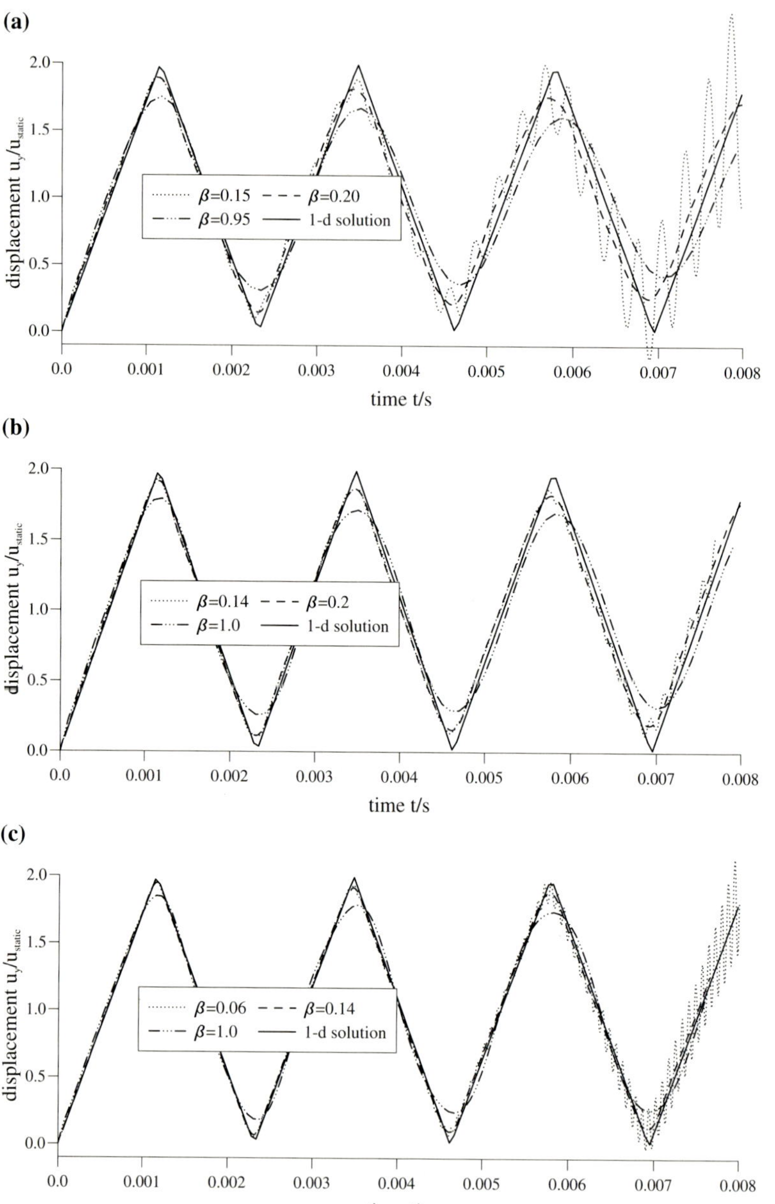

Fig. 4.7. Longitudinal displacement at the free of the rod versus time: Influence of time step size, i.e., different β **(a)** mesh 1 **(b)** mesh 2 **(c)** mesh 3

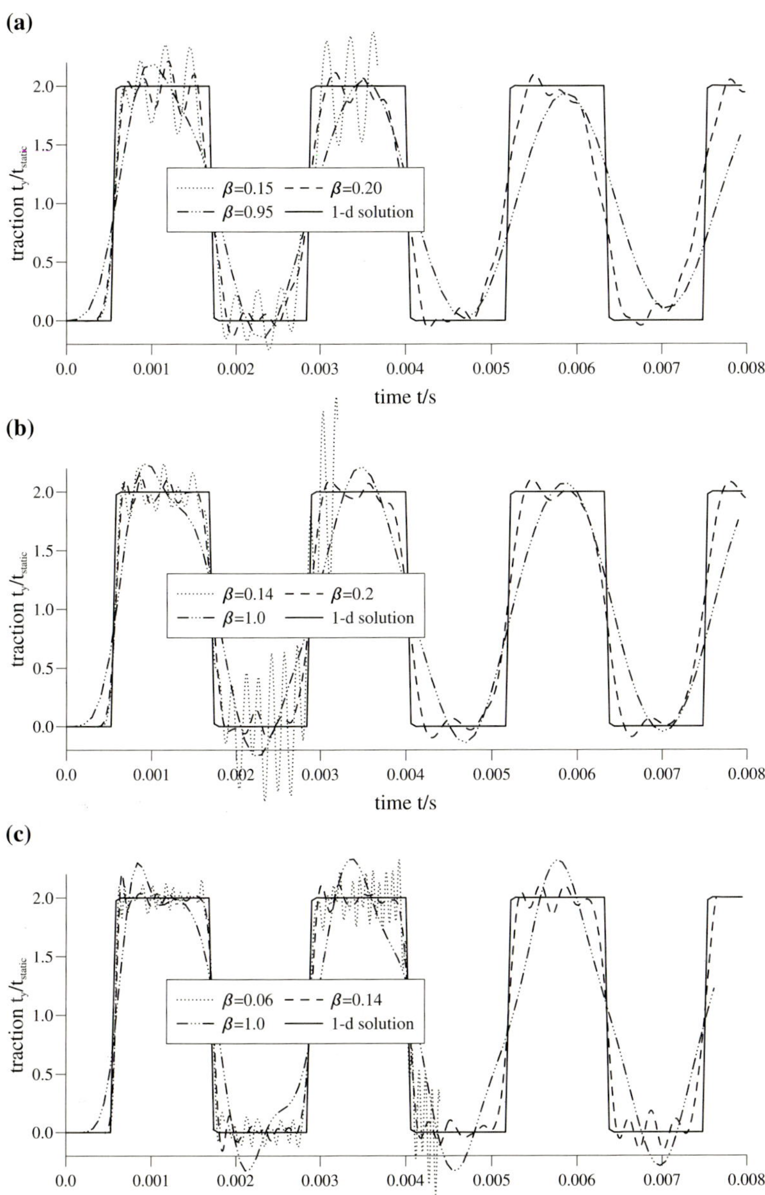

Fig. 4.8. Traction at the fixed end of the rod versus time: Influence of time step size, i.e., different β **(a)** mesh 1 **(b)** mesh 2 **(c)** mesh 3

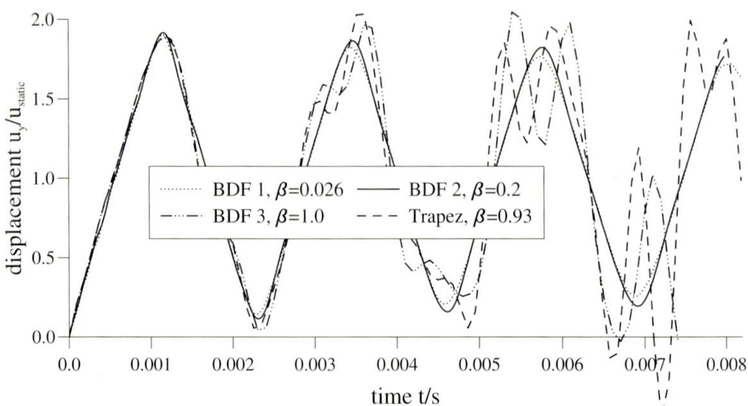

Fig. 4.9. Longitudinal displacement at the free of the rod versus time: Influence of the applied multistep method

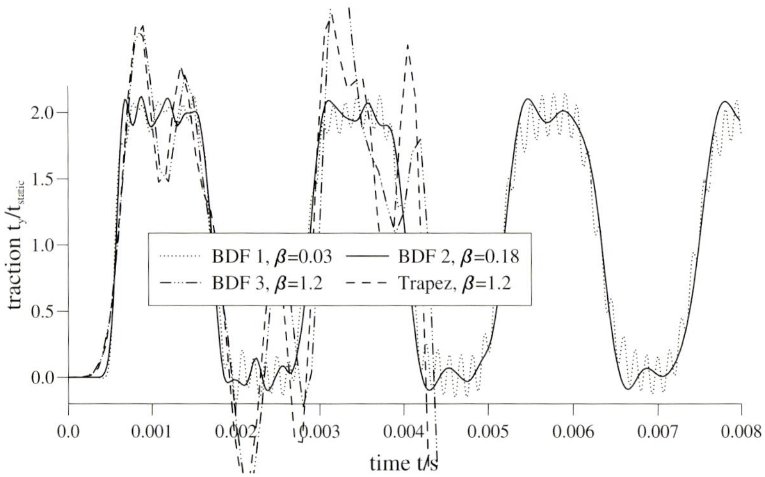

Fig. 4.10. Traction at the fixed end of the rod versus time: Influence of the applied multistep method

size for each of the different methods. The traction solution clearly indicates that the BDF 3 and the trapezoidal rule lead to not acceptable results, i.e., no stable time step size is found. As before, the unstable regions of the solutions are clipped. Contrary to BDF 3 and the trapezoidal rule, with BDF 1 and BDF 2 good solutions are achieved. This behavior was expected due to the results in Sect. 2.2. There, it was concluded that only A-stable methods which are also stable in infinity give satisfying results. Here, the results using the BDF 3 or the trapezoidal rule confirm this also for the elastodynamic boundary element formulation. But note, for a proper solution with the BDF 1 a very small time step size has to be taken to avoid numerical damping. Finally, it should be mentioned that the BDF 2 seems to be the best choice.

Therefore, in the following, all tests are performed with the BDF 2 as the underlying multistep method.

Remark on 2-d: Here, only 3-d results are presented. Indeed, the results can directly be transformed to the 2-d case. There is also a critical time step size for $\beta < 0.1$ which vanishes for very fine meshes. Concerning the multistep methods the same restriction to A-stable methods which are also stable in infinity has to be demanded.

4.3.2 Comparison with the "classical" time domain BE formulation

Above the proposed method was validated with a comparison to the 1-d analytical solution. Now, the method is compared to the classical boundary element method in time domain proposed by Mansur [126]. This method is widely used in elastodynamics or acoustic, but a time domain fundamental solution is necessary.

Compared to the method proposed here, the obvious advantage of the classical formulation concerns efficiency. In the classical formulation, the fundamental solutions are zero before the compression wave arrive at a distinct location r (causality of the solution) and in 3-d also when the slower shear wave had passed the location r. This condition represents that an elastic material has no memory in contrast to a viscoelastic material. Because causality and the second property are physical conditions the proposed formulation has to fulfill these conditions, too. Causality is implicit fulfilled by the summation over the integration weights in equation (4.20), i.e., the summation ends at n when $t = n\Delta t$. To validate the second condition an estimation of the behavior for large n of the integration weights (4.21) is necessary. Following the ideas presented in Sect. 2.2.1 to determine the integration weights analytically by a series expansion of the test functions (2.23) and (2.24), it can be shown that the order of magnitude of the integration weights are limited by

$$\omega_n^{ef}\left(\hat{U}_{ij},\mathbf{y},\Delta t\right) \approx \omega_n^{ef}\left(\hat{T}_{ij},\mathbf{y},\Delta t\right) \approx \mathrm{e}^{-1.5\frac{r_{max}}{c_2\Delta t}} \frac{\left(\frac{r_{max}}{c_2\Delta t}\right)^n}{n!} . \tag{4.26}$$

In equation (4.26), r_{max} is the maximum distance in the discretized body, i.e., the largest distance the slow shear wave c_2 has to travel. For the estimation (4.26) a BDF 2 as the underlying multistep is used, whereas other multistep methods will lead to other estimations. An upper limit $\bar{n}$ for calculating the integration weights can be estimated so that for all $n > \bar{n}$ the integration weights vanish in relation to weights $n < \bar{n}$.

The limit in the classical formulation $\bar{n} = \min\left(n, r_{max}/\left(c_2\Delta t\right) + 2\right)$ is smaller than this estimated $\bar{n}$. Therefore, the classical formulation is mostly more efficient than the proposed method. However, the efficiency can not be reduced to count the amount of necessary time steps, because both formulations have very different optimal time step sizes. As presented in the last section a $\beta \approx 0.2$ or even much smaller if the mesh is fine enough gives the best results and in the classical formulation $\beta \approx 0.8$ is a good choice. For all three meshes, in Fig. 4.11 the displacement at

(a)

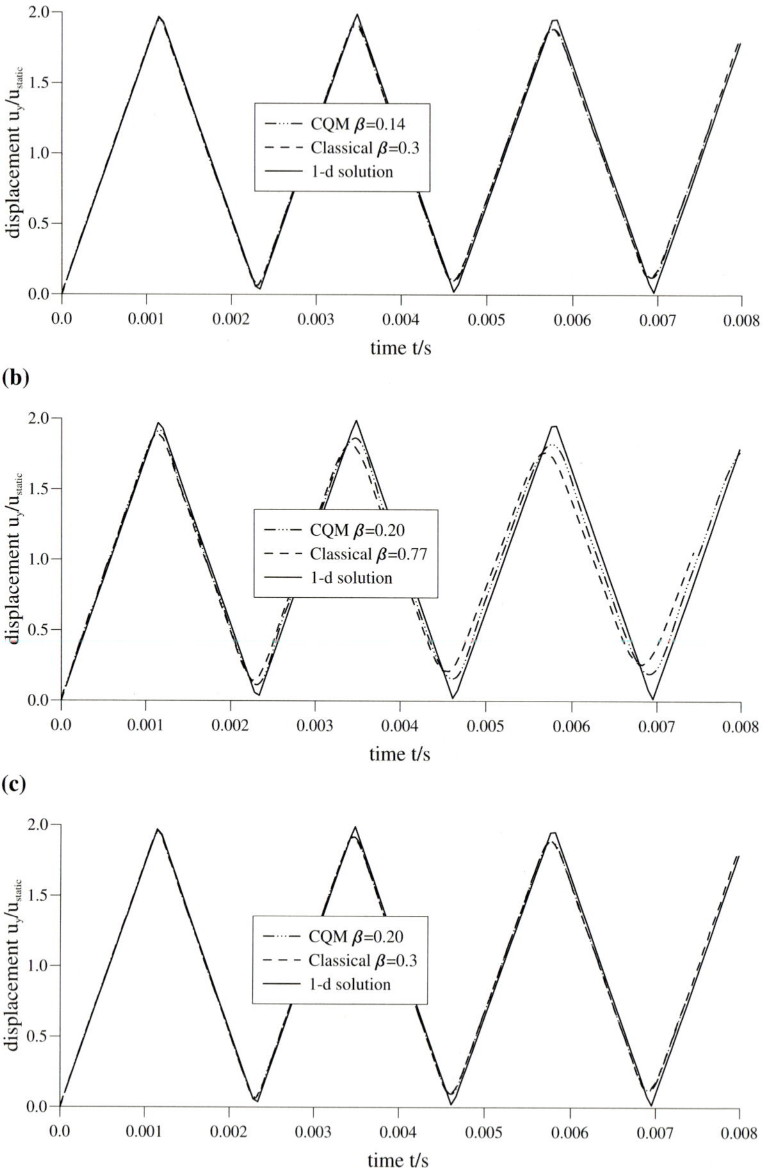

displacement u_y/u_{static}

CQM β=0.14
Classical β=0.3
1-d solution

time t/s

(b)

displacement u_y/u_{static}

CQM β=0.20
Classical β=0.77
1-d solution

time t/s

(c)

displacement u_y/u_{static}

CQM β=0.20
Classical β=0.3
1-d solution

time t/s

Fig. 4.11. Longitudinal displacement at the free of the rod versus time: Comparison of the proposed method with the classical method **(a)** mesh 1 **(b)** mesh 2 **(c)** mesh 3

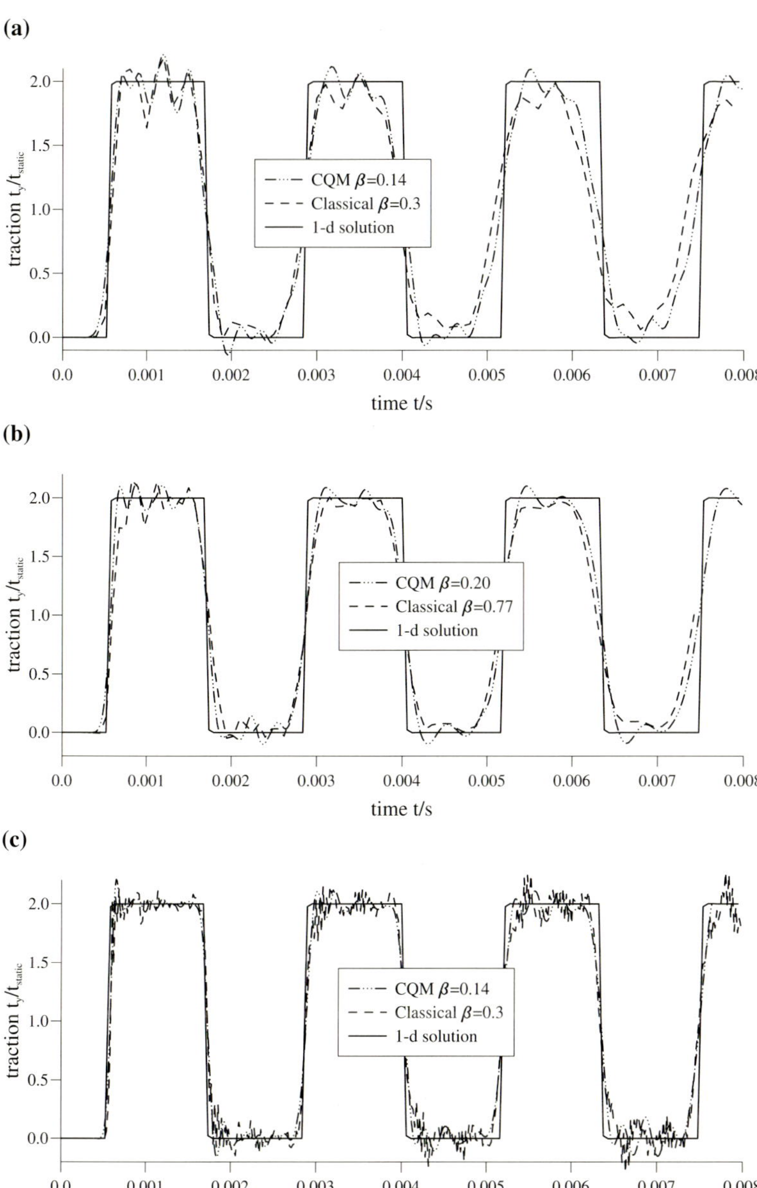

Fig. 4.12. Traction at the fixed end of the rod versus time: Comparison of the proposed method with the classical method **(a)** mesh 1 **(b)** mesh 2 **(c)** mesh 3

the free and, in Fig. 4.12, the traction at the fixed end of the rod, respectively, versus time is depicted using an optimal time step for each. Except for mesh 3, the proposed formulation is closer to the analytical solution than the classical one. For mesh 3 both methods lead to very good results. But note, for this mesh also with the classical formulation a small $\beta = 0.3$ is possible. For the coarse mesh 1 only the displacement solution of the new method can be accepted. The traction solution of both methods is not sufficient. Mesh 2 seems to be a good compromise between fine discretization, i.e., less CPU time and storage, and quality of the results.

Concluding this comparison, the classical formulation needs a finer spatial discretization than the new method. Presumably, this is caused due to the spatial integration. In the classical formulation Gaussian quadrature is used for a not continuous function. But, on the other hand the new formulation needs more CPU time. However, much smaller time step sizes are possible using the same mesh, if it is necessary, e.g., to approximate the time history of boundary conditions. Further, if domains with different materials, i.e., different wave velocities and, therefore, with different optimal time step sizes are considered the new method is advantageous due to the insensitivity concerning the time step size compared to the classical formulation.

5. Viscoelastodynamic boundary element formulation

Viscoelastic boundary element formulations are published for the quasi-static case (e.g., [175, 170, 44]), or in dynamics using a frequency or Laplace domain representation of the governing integral equation. These formulations are developed by applying the elastic-viscoelastic correspondence principle to the elastodynamic boundary element formulation, e.g., [115] for a frequency domain or [123] for a Laplace domain formulation.

Calculation of transient response, however, requires the inverse transformation. Since all numerical inversion formulas depend on a proper choice of their parameters [133], a direct evaluation in time domain seems to be preferable. But, formulations directly in time domain require the knowledge of viscoelastic fundamental solutions which are not yet known for the general viscoelastodynamic case. Only for a simple Maxwell model, a solution has been obtained analytically by Gaul and Schanz [90] and has been implemented in a boundary element formulation [165]. Based on the Laplace domain fundamental solutions which are numerically inverted within each time step a BE formulation in time domain was published by the same authors [92]. This formulation is very CPU time demanding, whereas using the convolution quadrature method of Lubich offers the more effective way [94].

First, constitutive equations for linear viscoelasticity are recalled and the well known elastic-viscoelastic correspondence principle is introduced.

5.1 Viscoelastic constitutive equation

The stress-strain relation of a linear isothermal viscoelastic material is given by the Stieltjes convolution [55] (see appendix A.2)

$$\sigma_{ij} = G_{ijkl} * \mathrm{d}\varepsilon_{kl} \ , \tag{5.1}$$

of $G_{ijkl}(t)$ the fourth order material tensor and $\mathrm{d}\varepsilon_{kl}(t)$ the first order derivative of the symmetric strain tensor $\varepsilon_{kl}(t)$. The fourth order material tensor $G_{ijkl}(t)$ fulfills the restrictions

$$G_{ijkl}(t) = 0 \qquad \text{for} \qquad -\infty < t < 0 \tag{5.2}$$

$$G_{ijkl}(t) = G_{jikl}(t) = G_{ijlk}(t) \ , \tag{5.3}$$

where the symmetry condition (5.3) is caused by the symmetry of the stress- and strain tensor. By decomposing the tensor $G_{ijkl}(t)$ the isotropic form of the viscoelastic stress-strain relation (5.1) is obtained. The most general isotropic representation of a fourth order tensor is

$$G_{ijkl}(t) = \frac{1}{3}\left[G_2(t) - G_1(t)\right]\delta_{ij}\delta_{kl} + \frac{1}{2}G_2(t)\left(\delta_{ik}\delta_{jl} + \delta_{il}\delta_{jk}\right) , \tag{5.4}$$

where $G_1(t)$ and $G_2(t)$ are independent relaxation functions and δ_{ij} is the Kronecker symbol. Following a procedure analogous to that in the elastic theory whereby the deviatoric components of both the stress tensor $s_{ij}(t)$ and the strain tensor $e_{ij}(t)$ are introduced, the constitutive equation (5.1) reduces to

$$s_{ij} = G_1 * \mathrm{d}e_{ij} \tag{5.5a}$$

$$\sigma_{ll} = G_2 * \mathrm{d}\varepsilon_{ll} \tag{5.5b}$$

for the deviatoric and the hydrostatic part, with

$$s_{ij} = \sigma_{ij} - \frac{1}{3}\delta_{ij}\sigma_{ll}, \quad s_{ii} = 0 \qquad \text{and} \tag{5.6a}$$

$$e_{ij} = \varepsilon_{ij} - \frac{1}{3}\delta_{ij}\varepsilon_{ll}, \quad e_{ii} = 0 \, . \tag{5.6b}$$

In a similar way, the constitutive equations can be formulated by creep functions [55].

The representation of the viscoelastic stress-strain relation as an integral equation is not the only possible form. An alternative formulation involving differential operators is also possible, and will be used in the following. The differential operator form of (5.5) may be written for the deviatoric and the hydrostatic part of the stress-strain relations

$$\sum_{k=0}^{N} p_k^D \frac{\mathrm{d}^k}{\mathrm{d}t^k} s_{ij} = \sum_{k=0}^{N} q_k^D \frac{\mathrm{d}^k}{\mathrm{d}t^k} e_{ij} \tag{5.7a}$$

$$\sum_{k=0}^{M} p_k^H \frac{\mathrm{d}^k}{\mathrm{d}t^k} \sigma_{ll} = \sum_{k=0}^{M} q_k^H \frac{\mathrm{d}^k}{\mathrm{d}t^k} \varepsilon_{ll} \, . \tag{5.7b}$$

The parameters p_k^D, q_k^D and N of the deviatoric part may differ from the parameter p_k^H, q_k^H and M of the hydrostatic part of the stress-strain relation which is in accordance with the two independent relaxation functions $G_1(t)$ and $G_2(t)$ of the integral representation (5.5). Applying the Laplace transform to equation (5.7) leads to

$$\sum_{k=0}^{N} p_k^D s^k \hat{s}_{ij} = \sum_{k=0}^{N} q_k^D s^k \hat{e}_{ij} \tag{5.8a}$$

$$\sum_{k=0}^{M} p_k^H s^k \hat{\sigma}_{ll} = \sum_{k=0}^{M} q_k^H s^k \hat{\varepsilon}_{ll} \, . \tag{5.8b}$$

In the transformed equation (5.8), it is assumed that the initial conditions of stresses and strains fulfill the relations

$$\begin{aligned}
&\sum_{r=k}^{N} p_r^D \frac{\mathrm{d}^{r-k}}{\mathrm{d}t^{r-k}} s_{ij}(0) = \sum_{r=k}^{N} q_r^D \frac{\mathrm{d}^{r-k}}{\mathrm{d}t^{r-k}} e_{ij}(0) \quad k = 1,2,\ldots,N \\
&\sum_{r=k}^{M} p_r^H \frac{\mathrm{d}^{r-k}}{\mathrm{d}t^{r-k}} \sigma_{ll}(0) = \sum_{r=k}^{M} q_r^H \frac{\mathrm{d}^{r-k}}{\mathrm{d}t^{r-k}} \varepsilon_{ll}(0) \quad k = 1,2,\ldots,M \,.
\end{aligned} \tag{5.9}$$

With these assumptions, the two representations of viscoelastic constitutive equations (5.5) and (5.7) are equivalent and consequently

$$s\hat{G}_1 = \frac{\sum\limits_{k=0}^{N} s^k q_k^D}{\sum\limits_{k=0}^{N} s^k p_k^D} \qquad s\hat{G}_2 = \frac{\sum\limits_{k=0}^{M} s^k q_k^H}{\sum\limits_{k=0}^{M} s^k p_k^H} \,. \tag{5.10}$$

Two different relaxation functions or, equivalent, different parameters in (5.7) is not only an academic case. For many materials, e.g., polymers, the behavior of the hydrostatic part is nearly elastic and the deviatoric part is viscoelastic. But, other materials, e.g., concrete, behave similar in both parts, i.e., the relaxation functions are the same [57].

The parameters or the relaxation functions have to be determined by curve fitting of measured data, where good agreement can be obtained by a relatively small number of parameters when fractional derivatives in the constitutive equations are introduced [176]. Furthermore, the fractional derivative concept ensures a causal behavior of non-causal rheological models, e.g., the Kelvin model [88]. Opposite to the very complicated various definitions of fractional derivatives in time domain, the Laplace transform reveals the useful result

$$\mathscr{L}\left\{\frac{\mathrm{d}^\alpha}{\mathrm{d}t^\alpha} x(t)\right\} = s^\alpha \mathscr{L}\{x(t)\} = s^\alpha \hat{x} \qquad \alpha \in \mathbb{C}. \tag{5.11}$$

For simplicity, in the above definition (5.11), vanishing initial conditions are assumed. With this assumption and definition (5.11), the constitutive equations for a generalized viscoelastic material in Laplace domain are

$$\sum_{k=0}^{N} s^{\alpha_k^D} p_k^D \hat{s}_{ij} = \sum_{k=0}^{N} s^{\beta_k^D} q_k^D \hat{e}_{ij} \tag{5.12a}$$

$$\sum_{k=0}^{M} s^{\alpha_k^H} p_k^H \hat{\sigma}_{ll} = \sum_{k=0}^{M} s^{\beta_k^H} q_k^H \hat{\varepsilon}_{ll} \,. \tag{5.12b}$$

Due to the independence of the deviatoric and the hydrostatic part of the 3-d constitutive equations, in each part a 1-d viscoelastic model can be used. The simplest model ensuring causal behavior is the three-parameter model, sometimes referred to as Kelvin-Voigt model (see Fig. 5.1). When the system is subjected to a step load,

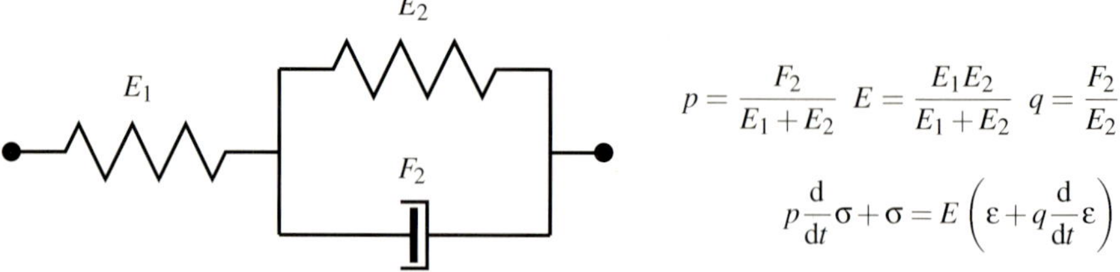

Fig. 5.1. One-dimensional rheological three-parameter model

it instantly deforms in an elastic state characterized by the spring constant E_1. As time progresses, the resistance offered by the dash-pot diminishes and the system softens. At large times, the apparent spring constant becomes $E = E_1 E_2 / (E_1 + E_2)$, which is smaller than the initial modulus E_1. The speed of the creep is regulated by the dashpot viscosity F_2. A characteristic time scale for the creep can be defined as $q = F_2 / E_2$. The appropriate constitutive relation is also given in Fig. 5.1.

Inserting fractional derivatives in the Laplace transformed constitutive equation in Fig. 5.1 the generalized model

$$(1 + p s^{\alpha}) \hat{\sigma} = E \left(1 + q s^{\beta} \right) \hat{\varepsilon} \tag{5.13}$$

depending on five parameters is found. Modeling a solid material with this constitutive equation requires a finite initial modulus $G_1 \ (t = 0)$. This initial modulus can be deduced in Laplace domain using the *initial value theorem* [68]

$$G_1 (0) = \lim_{s \to \infty} s \hat{G}_1 (s) = \lim_{s \to \infty} E \frac{1 + q s^{\beta}}{1 + p s^{\alpha}} = \lim_{s \to \infty} E \frac{\frac{1}{s^{\alpha}} + q s^{\beta - \alpha}}{\frac{1}{s^{\alpha}} + p} \ . \tag{5.14}$$

A finite value is achieved only by choosing $\alpha = \beta$. Other values lead to an infinite or a zero initial modulus. A plausible model of the viscoelastic phenomenon should also predict nonnegative internal work and a nonnegative rate of energy dissipation [17]. Therefore, the parameters are constrained by [91]

$$0 < E \qquad 0 \leq p < q \qquad 0 < \alpha = \beta < 2 \ . \tag{5.15}$$

Applying this one-dimensional model to three dimensions, the constitutive equations in Laplace domain are

$$\left(1 + p^D s^{\alpha^D} \right) \hat{s}_{ij} = 2G \hat{e}_{ij} \left(1 + q^D s^{\alpha^D} \right) \tag{5.16a}$$

$$\left(1 + p^H s^{\alpha^H} \right) \hat{\sigma}_{ll} = 3K \hat{\varepsilon}_{ll} \left(1 + q^H s^{\alpha^H} \right) \tag{5.16b}$$

with the elastic compression modulus K and the elastic shear modulus G.

Comparing the viscoelastic constitutive equations (5.16) with Hook's law, the elastic-viscoelastic correspondence principle is obtained

$$2G \Longleftrightarrow 2G \frac{1 + q^D s^{\alpha^D}}{1 + p^D s^{\alpha^D}} \qquad 3K \Longleftrightarrow 3K \frac{1 + q^H s^{\alpha^H}}{1 + p^H s^{\alpha^H}} \ . \tag{5.17}$$

This means, every elastodynamic solution of a distinct problem can be converted to the solution of the related viscoelastic problem by inserting the correspondence (5.17) [55]. Mostly, and also in the following, instead of the compression modulus and the shear modulus, Young's modulus E and Poisson's ratio ν are used. Therefore, the elastic-viscoelastic correspondence for these two material values are introduced

$$E \Leftrightarrow \frac{3E\left(1+q^H s^{\alpha^H}\right)\left(1+q^D s^{\alpha^D}\right)}{2(1+\nu)\left(1+q^H s^{\alpha^H}\right)\left(1+p^D s^{\alpha^D}\right)+(1-2\nu)\left(1+q^D s^{\alpha^D}\right)\left(1+p^H s^{\alpha^H}\right)} \tag{5.18}$$

$$\nu \Leftrightarrow \frac{(1+\nu)\left(1+q^H s^{\alpha^H}\right)\left(1+p^D s^{\alpha^D}\right)-(1-2\nu)\left(1+q^D s^{\alpha^D}\right)\left(1+p^H s^{\alpha^H}\right)}{2(1+\nu)\left(1+q^H s^{\alpha^H}\right)\left(1+p^D s^{\alpha^D}\right)+(1-2\nu)\left(1+q^D s^{\alpha^D}\right)\left(1+p^H s^{\alpha^H}\right)} . \tag{5.19}$$

In equation (5.19), the dependence from the Laplace parameter s is clearly observed, which corresponds to a time-dependent Poisson's ratio. If, however, the relation

$$\left(1+q^H s^{\alpha^H}\right)\left(1+p^D s^{\alpha^D}\right)=\left(1+q^D s^{\alpha^D}\right)\left(1+p^H s^{\alpha^H}\right) , \tag{5.20}$$

is satisfied, i.e., the deviatoric and the hydrostatic part of the stress-strain relation behaves similar, the influence of the Laplace parameter s vanishes and, finally, yields a time invariant Poisson's ratio equal to the elastic case [55].

5.2 Boundary integral equation

The equation of motion for a viscoelastic continuum is achieved by inserting the constitutive equation (5.1) in the dynamic equilibrium (4.3)

$$\left(G_{ijkl} * \mathrm{d}\varepsilon_{kl}\right)_{,j} + b_i = \rho \ddot{u}_i . \tag{5.21}$$

Assuming homogeneity, isotropy (5.4), and the linear strain-displacement relation (4.2) yields another representation of the equation of motion, more convenient for wave propagation problems, the viscoelastic counterpart to the Lamé equation

$$\frac{G_1}{2} * \mathrm{d}u_{i,kk} + \frac{1}{3}\left(2G_2 + G_1\right) * \mathrm{d}u_{k,ki} + b_i = \rho \ddot{u}_i . \tag{5.22}$$

There are two possibilities to obtain an integral equation for (5.22). One is the weighted residual technique as used for elastodynamics in Sect. 4.1 and the other uses a reciprocal work theorem, e.g., in the elastic case Betti's theorem. Here, a viscoelastic reciprocal work theorem will be used. Gurtin and Sternberg [102] have developed such a theorem for the quasi-static case which has been extended by de Hoop [66] to dynamics. A detailed derivation of this dynamic reciprocal work theorem for viscoelastodynamics is given in the following.

Starting point is the equivalence of

$$\int_\Omega \sigma_{ij} * \varepsilon'_{ij} \mathrm{d}\Omega = \int_\Omega \sigma'_{ij} * \varepsilon_{ij} \mathrm{d}\Omega \,, \tag{5.23}$$

where $[\mathbf{u},\varepsilon,\sigma]$ and $[\mathbf{u}',\varepsilon',\sigma']$ are two viscoelastic states as defined by Gurtin and Sternberg [102]. The equivalence (5.23) can easily be proven using the properties of the Stieltjes convolution [102] and the constitutive equation

$$\sigma_{ij} * \varepsilon'_{ij} = G_{ijkl} * \mathrm{d}\varepsilon_{kl} * \varepsilon'_{ij} = \mathrm{d}G_{ijkl} * \varepsilon_{kl} * \varepsilon'_{ij} = \mathrm{d}G_{ijkl} * \varepsilon'_{ij} * \varepsilon_{kl} = \sigma'_{kl} * \varepsilon_{kl} \,. \tag{5.24}$$

This equivalence is valid in quasi-statics [102] as well as in dynamics.

Inserting the linear stress-strain relation (4.2) in the left integral of equation (5.23) and a subsequent integration by parts results in

$$\int_\Omega \sigma_{ij} * \varepsilon'_{ij} \mathrm{d}\Omega = \int_\Omega \sigma_{ij} * u'_{i,j} \mathrm{d}\Omega = \int_\Omega \left(\sigma_{ij} * u'_i\right)_{,j} \mathrm{d}\Omega - \int_\Omega \sigma_{ij,j} * u'_i \mathrm{d}\Omega \,. \tag{5.25}$$

Then, the divergence theorem $\int_\Omega \left(\sigma_{ij} * u'_i\right)_{,j} \mathrm{d}\Omega = \int_\Gamma \sigma_{ij} n_j * u'_i \mathrm{d}\Gamma$, the definition of a stress vector (Cauchy's theorem) $t_i = \sigma_{ij} n_j$ (4.1), and the equation of motion (4.2) lead to

$$\int_\Omega \sigma_{ij} * \varepsilon'_{ij} \mathrm{d}\Omega = \int_\Gamma t_i * u'_i \mathrm{d}\Gamma + \int_\Omega b_i * u'_i \mathrm{d}\Omega - \int_\Omega \ddot{u}_i * u'_i \mathrm{d}\Omega \,. \tag{5.26}$$

For the integral on the right hand side of equation (5.23), a similar relation can be deduced following the same operations simply by exchanging the viscoelastic states. Finally, expressing the equivalence (5.23) by the equation (5.26) and the analogous equation for the exchanged states, the reciprocal work theorem for two viscoelastic states $[\mathbf{u},\varepsilon,\sigma]$ and $[\mathbf{u}',\varepsilon',\sigma']$

$$\int_\Gamma t_i * u'_i \mathrm{d}\Gamma + \int_\Omega b_i * u'_i \mathrm{d}\Omega = \int_\Gamma t'_i * u_i \mathrm{d}\Gamma + \int_\Omega b'_i * u_i \mathrm{d}\Omega \tag{5.27}$$

is obtained. In the reciprocal theorem vanishing initial conditions for the displacements $u_i(\mathbf{x},0)$, $u'_i(\mathbf{x},0)$ and velocities $\dot{u}_i(\mathbf{x},0)$, $\dot{u}'_i(\mathbf{x},0)$ are assumed. Therefore, the inertia terms cancel one another because of the relation

$$\ddot{u}_i * u'_i = u_i * \ddot{u}'_i \quad \text{for} \quad u_i(\mathbf{x},0) = 0,\ \dot{u}_i(\mathbf{x},0) = 0,\ u'_i(\mathbf{x},0) = 0,\ \dot{u}'_i(\mathbf{x},0) = 0 \,. \tag{5.28}$$

This relation can be shown by two partial integrations with respect to time (see (4.10)).

Now, let the viscoelastic state $[\mathbf{u}',\varepsilon',\sigma']$ in the reciprocal work theorem (5.27) be defined in the full space and subjected to a point load at the field point $\mathbf{y}$ at time τ in the direction x_j, i.e., the inhomogeneity in the equation of motion is $b'_i = \delta(\mathbf{y}-\mathbf{x})\,\delta(t-\tau)\,\delta_{ij}$. The solutions of the equation of motion due to such a load are the displacement $U_{ij}(\mathbf{x},\mathbf{y},t-\tau)$ and the traction fundamental solution $T_{ij}(\mathbf{x},\mathbf{y},t-\tau)$. Inserting this viscoelastic state in the reciprocal work theorem (5.27) leads to

$$u_j(\mathbf{y},t) = \int_0^t \int_\Gamma \left[U_{ij}(\mathbf{x},\mathbf{y},t-\tau)\,t_i(\mathbf{x},\tau) - T_{ij}(\mathbf{x},\mathbf{y},t-\tau)\,u_i(\mathbf{x},\tau)\right] \mathrm{d}\Gamma\,\mathrm{d}\tau \quad \mathbf{y}\in\Omega\,, \tag{5.29}$$

when vanishing body forces $b_i = 0$ of the other state $[\mathbf{u},\varepsilon,\sigma]$ are assumed. Of course, non vanishing body forces can be treated similar to the elastodynamic case, but for simplicity the assumption of vanishing body forces is chosen here. The integral free term results from the last integral in (5.27) due to the properties of the generalized function $b'_i = \delta(\mathbf{x}-\mathbf{y})\,\delta(t-\tau)\,\delta_{ij}$.

As indicated in equation (5.29), the boundary integral equation is only valid for points $\mathbf{y}$ lying in the domain Ω. For $\mathbf{y}$ approaching the boundary Γ, i.e., $\mathbf{y}\to\mathbf{x}$ or $r\to 0$ with $r_i = x_i - y_i$ and $r=\sqrt{r_i r_i}$, the Laplace transformed fundamental solutions become singular (see appendix B.1 for the explicit expressions). To study this singular behavior, the power series expansion of the exponential function e^{-rs/c_i} is inserted in the fundamental solutions and subsequently the result is rearranged with ascending power of r. For the displacement solution we have

$$\hat{U}_{ij}(\mathbf{x},\mathbf{y},s) = \underbrace{\frac{1+\nu(s)}{8\pi E(s)\,(1-\nu(s))}\left\{r_{,i}r_{,j} + \delta_{ij}\,(3-4\nu(s))\right\}\frac{1}{r}}_{\text{static solution}} + \mathcal{O}\left(r^0\right) \tag{5.30}$$

and for the tractions

$$\hat{T}_{ij}(\mathbf{x},\mathbf{y},s) = \underbrace{-\frac{\left[(1-2\nu(s))\,\delta_{ij} + 3r_{,i}r_{,j}\right]r_{,n} - (1-2\nu(s))\,(r_{,j}n_i - r_{,i}n_j)}{8\pi\,(1-\nu(s))\;r^2}}_{\text{static solution}} + \mathcal{O}\left(r^0\right) \tag{5.31}$$

where $E(s)$ and $\nu(s)$ are abbreviations for the lengthy expressions on the right hand side of the equations (5.18) and (5.19), respectively. The series expansions of the displacement (5.30) and traction (5.31) shows a weak $\mathcal{O}\left(r^{-1}\right)$ and a strong $\mathcal{O}\left(r^{-2}\right)$ singularity, respectively. The under-braced expressions are nearly identical to the elastostatic fundamental solutions in (4.15) and (4.16), with the difference that instead of the elastic material parameters E and ν, here, the expressions $E(s)$ and $\nu(s)$ are dependent on the Laplace parameter s. But, this dependence does not change the singular behavior with respect to r. As a consequence, the time domain fundamental solutions must have the same singularities as in the transformed domain. Therefore, the limit of $\mathbf{y}$ to the boundary Γ in equation (5.29) can be performed analogous to the elastostatic case (see Sect. 4.1) and leads to the viscoelastic boundary integral equation

$$\int_0^t c_{ij}(\mathbf{y},t-\tau)\,u_i(\mathbf{y},\tau)\,\mathrm{d}\tau = \int_0^t \int_\Gamma U_{ij}(\mathbf{x},\mathbf{y},t-\tau)\,t_i(\mathbf{x},\tau)\,\mathrm{d}\Gamma\mathrm{d}\tau - \int_0^t \oint_\Gamma T_{ij}(\mathbf{x},\mathbf{y},t-\tau)\,u_i(\mathbf{x},\tau)\,\mathrm{d}\Gamma\mathrm{d}\tau \quad \mathbf{y}\in\Omega\cup\Gamma\,. \tag{5.32}$$

Due to the time dependence of Poisson's ratio $\nu(t)$, now, in general $c_{ij}(t)$ is time-dependent, too. But, the integral free terms $c_{ij}(t)$ are independent of time and equal to the elastostatic case, if Poisson's ratio is independent of time. This is true, if equation (5.20) is valid, i.e., the hydrostatic and the deviatoric part of the stress-strain relation behaves similar.

Further on, as in the elastostatic case, the first integral in equation (5.32) is weakly singular, and the second integral has to be defined in the sense of a Cauchy principal value, i.e., the integral is strongly singular.

5.3 Boundary element formulation

The numerical implementation of equation (5.32) follows the same procedure as in the elastodynamic case (Sect. 4.2). After discretizing the boundary Γ with E elements and approximating the displacement $u_i(\mathbf{x},t)$ and traction $t_i(t,\mathbf{x})$ by polynomial shape functions (4.18) the integral equation (5.32) reads

$$c_{ij}(\mathbf{y},t)*u_i(\mathbf{y},t) = \sum_{e=1}^{E}\sum_{f=1}^{F}\left[\int_\Gamma U_{ij}(\mathbf{x},\mathbf{y},t)N_e^f(\mathbf{x})\,\mathrm{d}\Gamma * t_i^{ef}(t) - \oint_\Gamma T_{ij}(\mathbf{x},\mathbf{y},t)N_e^f(\mathbf{x})\,\mathrm{d}\Gamma * u_i^{ef}(t)\right]. \tag{5.33}$$

In the next step a time discretization is necessary. In the elastodynamic case there are in general two alternatives: the classical method using time dependent fundamental solutions and the method based on the convolution quadrature method. Here, in the viscoelastic case there are no time-dependent fundamental solutions available, therefore, an analogous method to the classical method is not possible. Hence, the convolution quadrature method is the only effective choice compared to the possibility to invert the fundamental solutions at every collocation point in every time step as proposed in [167].

Hence, after dividing the time t in N intervals of equal duration Δt so that $t = N\Delta t$, the convolution integral between the fundamental solutions $U_{ij}(\mathbf{x},\mathbf{y},t)$ or $T_{ij}(\mathbf{x},\mathbf{y},t)$ and the nodal values $t_i^{ef}(t)$ or $u_i^{ef}(t)$, respectively, is performed by the convolution quadrature method, i.e., the quadrature formula (2.18) is applied to the integral equation (5.33). This results in the following boundary element time-stepping formulation for $n = 0,1,\ldots,N$

$$\begin{aligned}&\sum_{k=0}^{n} \omega_{n-k}\left(\hat{c}_{ij},\mathbf{y},\Delta t\right) u_i\left(\mathbf{y},k\Delta t\right)\\&\quad=\sum_{e=1}^{E}\sum_{f=1}^{F}\sum_{k=0}^{n}\left[\omega_{n-k}\left(\hat{U}_{ij},\mathbf{y},\Delta t\right) t_i^{ef}\left(k\Delta t\right)-\omega_{n-k}\left(\hat{T}_{ij},\mathbf{y},\Delta t\right) u_i^{ef}\left(k\Delta t\right)\right]\end{aligned} \tag{5.34}$$

with the integration weights corresponding to equation (2.22)

$$\omega_n\left(\hat{c}_{ij},\mathbf{y},\Delta t\right)=\frac{\mathscr{R}^{-n}}{L}\sum_{\ell=0}^{L-1}\hat{c}_{ij}\left(\mathbf{y},\frac{\gamma\left(\mathscr{R}\mathrm{e}^{-\mathrm{i}\ell\frac{2\pi}{L}}\right)}{\Delta t}\right)\mathrm{e}^{-\mathrm{i}n\ell\frac{2\pi}{L}}\,, \tag{5.35a}$$

$$\omega_n\left(\hat{U}_{ij},\mathbf{y},\Delta t\right)=\frac{\mathscr{R}^{-n}}{L}\sum_{\ell=0}^{L-1}\int_{\Gamma}\hat{U}_{ij}\left(\mathbf{x},\mathbf{y},\frac{\gamma\left(\mathscr{R}\mathrm{e}^{-\mathrm{i}\ell\frac{2\pi}{L}}\right)}{\Delta t}\right)N_e^f\left(\mathbf{x}\right)\mathrm{d}\Gamma\;\mathrm{e}^{-\mathrm{i}n\ell\frac{2\pi}{L}}\,, \tag{5.35b}$$

$$\omega_n\left(\hat{T}_{ij},\mathbf{y},\Delta t\right)=\frac{\mathscr{R}^{-n}}{L}\sum_{\ell=0}^{L-1}\oint_{\Gamma}\hat{T}_{ij}\left(\mathbf{x},\mathbf{y},\frac{\gamma\left(\mathscr{R}\mathrm{e}^{-\mathrm{i}\ell\frac{2\pi}{L}}\right)}{\Delta t}\right)N_e^f\left(\mathbf{x}\right)\mathrm{d}\Gamma\,\mathrm{e}^{-\mathrm{i}n\ell\frac{2\pi}{L}}\,. \tag{5.35c}$$

Obviously, although using only the Laplace transformed fundamental solutions $\hat{U}_{ij}(\mathbf{x},\mathbf{y},s)$ and $\hat{T}_{ij}(\mathbf{x},\mathbf{y},s)$, a time stepping procedure directly in time domain is achieved, i.e., a viscoelastic boundary element formulation in time domain is obtained without the knowledge of the time dependent fundamental solutions. Note, in general, as mentioned above, the integral free terms $c_{ij}(t)$ are time-dependent in viscoelasticity and, therefore, have to be convoluted with the displacements. This time dependence of the integral free terms $c_{ij}(t)$ is shown in Fig. 5.2. Exemplary,

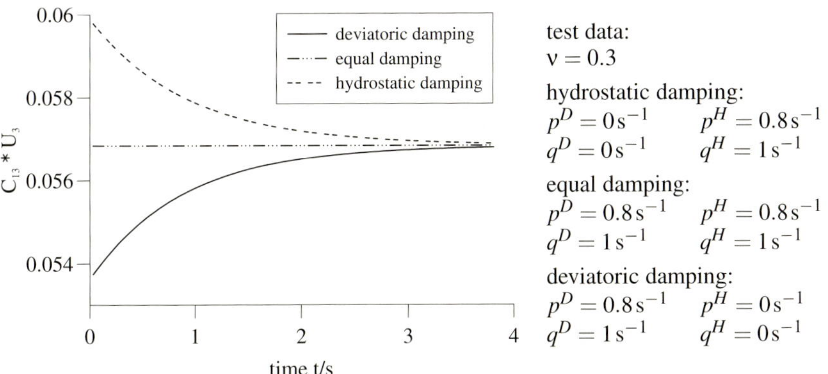

Fig. 5.2. Convolution $c_{13} * u_3$ versus time for a viscoelastic material

the coefficient c_{13} of a rectangular corner $\hat{c}_{13}(s) = 1/\left(8\pi\left(1-\nu(s)\right)\right)$ is convoluted with a unit step function $u_3 = H(t) \cdot 1\,\text{m}$. The related elastic Poisson's ratio in this test is $\nu = 0.3$ leading to $c_{13} = 0.0568$, and a BDF 2 as the underlying multistep method is used. Three cases are considered in Fig. 5.2:

i) hydrostatic damping – the deviatoric part of the stress-strain relation has an elastic behavior,
ii) equal damping mechanisms for the deviatoric and hydrostatic part of the stress-strain relation, and
iii) deviatoric damping – the hydrostatic part of the stress-strain relation has an elastic behavior.

The initial values of the three cases are quite different. With increasing time a relaxation or a creep behavior of the results is observed. Equal damping mechanisms results, as expected, in a constant value of c_{13} equal to the elastic case.

Further, the spatial integration has to be discussed. Because the integral over each boundary element involves only the Laplace transformed fundamental solution (see equations (5.35b) and (5.35c)), in case of regular integrals a standard Gaussian quadrature rule can be used. The singular parts of the fundamental solutions are similar to the elastostatic kernels as shown in equations (5.30) and (5.31). Due to this, the same regularization methods can be applied, i.e., the weakly singular integrals in (5.35b) are regularized by coordinate transformation and the strongly singular integrals in (5.35c) by the method suggested by Guiggiani and Gigante [101].

The next step is identical to the elastodynamic case. In order to arrive in equation (5.34) at systems of algebraic equations, collocation is used at every node of the shape functions $N_e^f(\mathbf{x})$. Then, as in elastodynamics, due to the property of the integration weights ω_{n-k}^{ef} in equation (5.34), they are only dependent on the difference $n-k$ not on n, a similar recursion formula to (4.22) can be established ($m = n-k$)

$$\omega_0(\mathbf{C})\,\mathbf{d}^n = \omega_0(\mathbf{D})\,\bar{\mathbf{d}}^n + \sum_{m=1}^{n}\left(\omega_m(\mathbf{U})\,\mathbf{t}^{n-m} - \omega_m(\mathbf{T})\,\mathbf{u}^{n-m}\right) \qquad n = 1,2,\dots,N\,. \tag{5.36}$$

The matrices and vectors are defined as in equation (4.22). Finally, a direct equation solver is also in the viscoelastic case applied.

5.4 Validation of the method and parameter study

The propagation of waves in a 3-d viscoelastic continuum has been calculated by the presented viscoelastic boundary element formulation. Because the fundamental solutions are the only difference between the elasto- and viscoelastodynamic formulation and, however, these fundamental solutions are very similar to the elastodynamic ones, the same behavior as observed in Sect. 4.3 is expected. Therefore, the comparison of the 1-d solution with the proposed method applied to a 3-d rod is briefly described with respect to the influence of the spatial discretization and the time step size. The examination concerning different multistep methods is skipped, because no differences to the elastodynamic case have been observed. Therefore, here, a BDF 2 as the underlying multistep method is used. More interesting are studies concerning the influence of the viscoelastic parameters using either the 3-d rod and in the second part of this section, considering an elastic concrete foundation slab on a viscoelastic half space.

As in Chap. 4, for all tests the parameters of the convolution quadrature are chosen as suggested by Lubich [119]: $L = N$ and $\mathcal{R}^N = \sqrt{\varepsilon}$ with $\varepsilon = 10^{-10}$ (see also Sect. 2.2). The parameter β introduced in equation (4.25) is now calculated using the viscoelastic compression wave velocity c_1 defined with the initial moduli (see appendix B.1.1 equation (B.5)) and the characteristic element length r_e as described in Sect. 4.3.

5.4.1 Three-dimensional rod

The three-dimensional rod considered is fixed on one end, and excited by a pressure jump according to a unit step function $t_y(\mathbf{x},t) = 1\,\mathrm{N/m^2}\,H(t)$ on the other free end (Fig. 5.3). The remaining surfaces are traction free. Linear spatial shape functions on triangles are used. The material data selected are measured in a resonance test with a perspex (PMMA) strip[1]. However, for the comparison with a 1-d solution Poisson's ratio is set $\nu = 0$. This 1-d solution is obtained with the elastic-viscoelastic correspondence principle applied to the known solution of the elastic rod for the displacement (4.23) and traction (4.24), respectively [94]. Further, all results are

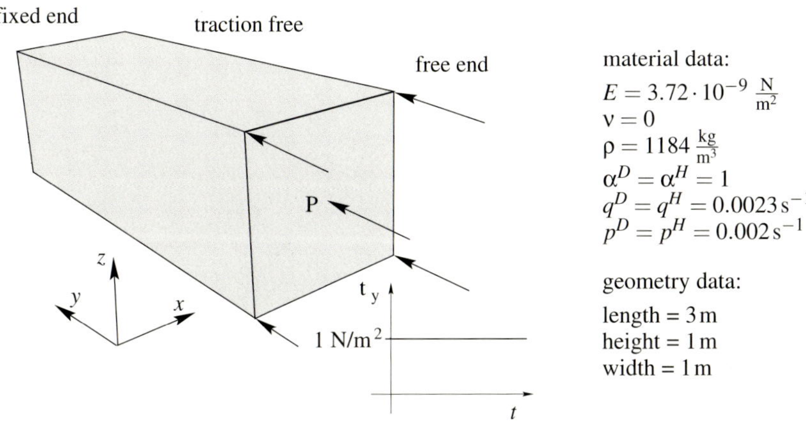

Fig. 5.3. Step function excitation of a free–fixed rod

normalized by the static values, i.e., the displacements by $u_{Static} = 8.065 \cdot 10^{-10}\,\mathrm{m}$ and the tractions by $t_{Static} = -1\,\mathrm{N/m^2}$, respectively. Static means here the results of

[1] The data are measured at the Institute of Technical Mechanics, Technical University Braunschweig

an elastostatic calculation using the Young's modulus and the Poisson's ratio given in Fig. 5.3.

First, the influence of the spatial discretization is studied using the same three meshes as in the previous test (see Fig. 4.4). The displacement at the midpoint of the free end (point P) and the traction at the midpoint of the fixed end in longitudinal direction is plotted versus time in Figs. 5.4 and 5.5, respectively. The results for

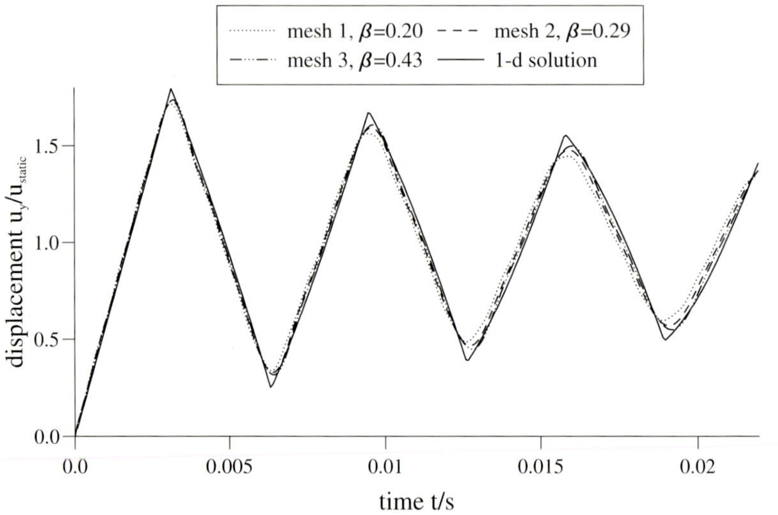

Fig. 5.4. Longitudinal displacement at the free end of the rod versus time: Influence of mesh size

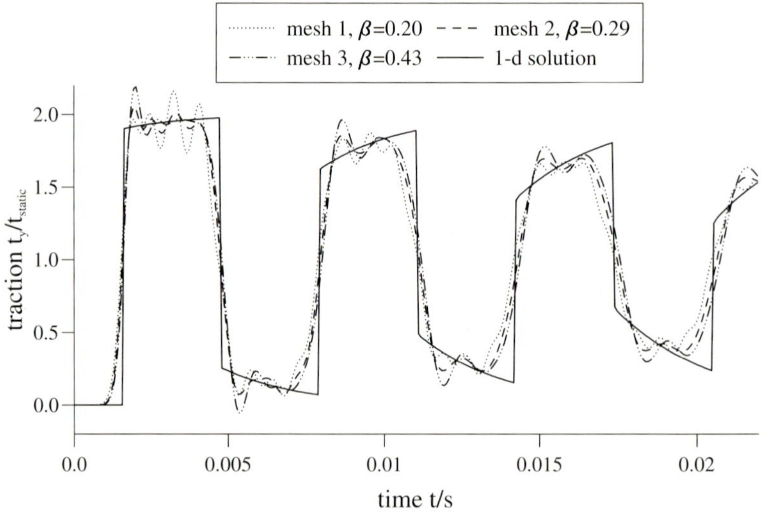

Fig. 5.5. Traction at the fixed end of the rod versus time: Influence of mesh size

the displacements as well as for the tractions are in good agreement with the 1-d solution for all meshes, even though the same time step size Δt is used leading to different values of β, i.e., only for mesh 1 the time step size is optimal. But even suboptimal time step sizes for mesh 2 and 3 yield satisfactory displacement results. In contrast, the traction solution requires mesh 3 or finer discretizations.

In Fig. 4.7, the influence of time step size on the elastodynamic formulation was studied. There, contrary to the classical formulation, a insensitivity concerning the time step size was observed if the mesh size is fine enough and $\beta < 1$ is regarded. The results for the viscoelastic formulation in Figs. 5.4 and 5.5 show a same tendency. Therefore, next, this will be investigated.

For mesh 2, the normalized displacement at point P is depicted in Fig. 5.6 for different β. Additionally, in Fig. 5.7, the traction at the support is given. The traction

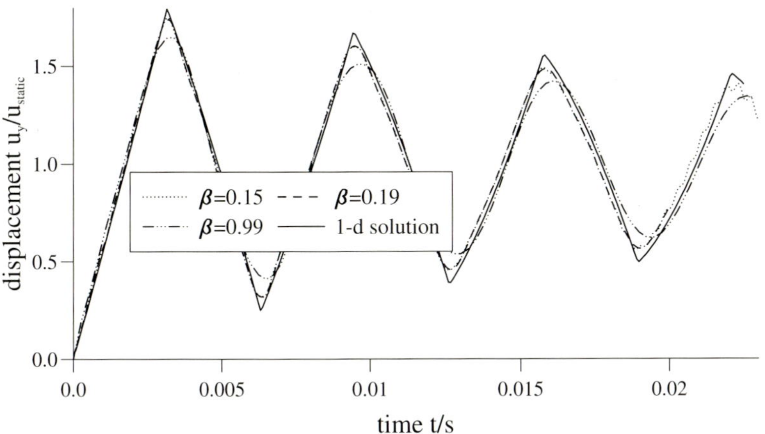

Fig. 5.6. Longitudinal displacement at the free end versus time: Influence of time step size, i.e., different β

solution indicates a critical time step size, corresponding to $\beta \lesssim 0.16$, below which the results are unstable. Stable displacement results deviate only slightly from the 1-d solution. However, the traction solution is for $\beta = 0.99$ and large t only a coarse approximation of the 1-d solution. Hence, mesh 3 or finer discretizations are needed to yield good results for the traction. This is in accordance with the experiences with the elastodynamic formulation. These experiences and the tests for the viscoelastic case give reason to the conclusion, that for a fine enough mesh the results are not dependent on the time step size, if $\beta < 1$ is regarded.

After these tests concerning the spatial and temporal discretization, now, the influence of the viscosity is considered. In Fig. 5.8, the longitudinal displacement at point P is plotted versus time for different values of $q^D = q^H = q$, with constant $p^D = p^H$ and $\alpha^D = \alpha^H$. In the used constitutive equation (5.16) the ratio

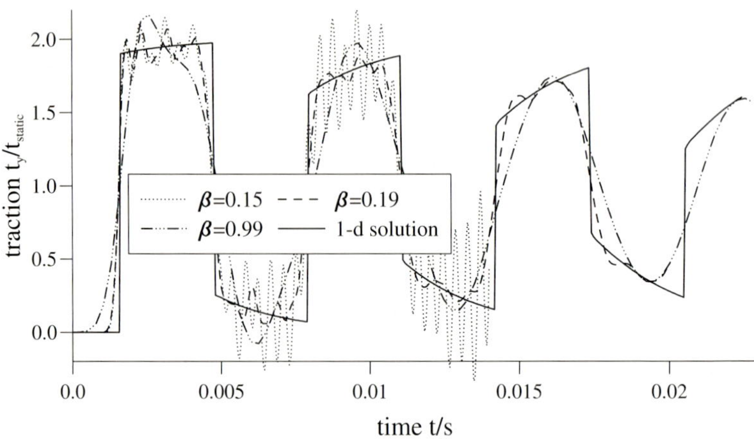

Fig. 5.7. Traction at the fixed end versus time: Influence of time step size, i.e., different β

q^D/p^D and q^H/p^H, respectively, characterize the damping. An increasing value of this ratio models an increasing damping. This becomes obvious in Fig. 5.8, where larger values of q, i.e., the ratio q/p becomes also larger, cause higher damping. The viscoelastic wave velocities also increase, because the material is stiffened due to the viscosity. This is also observed in Fig. 5.8. One remark should be added for $q = 0.002\,\mathrm{s}^{-1}$. In this case the ratio $q/p = 1$, i.e., no damping occur. Looking on the elastic-viscoelastic correspondence principle (5.17) it is obvious that this ratio represents the elastic case. The graph for $q = 0.002\,\mathrm{s}^{-1}$ would cover a graph of

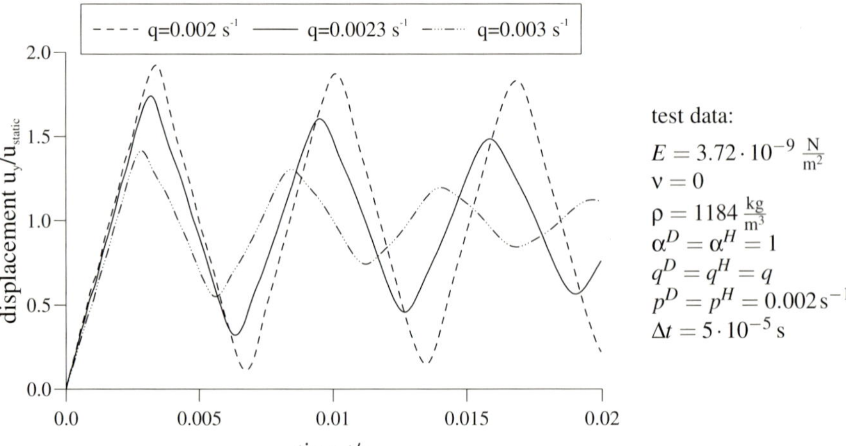

Fig. 5.8. Longitudinal displacement at the free end versus time: Influence of viscosity q

the corresponding elastodynamic solution and is, therefore, skipped here, i.e., the theoretical equivalence is confirmed by the numerical result.

Furthermore, the effect of the fractional order α is of interest. In Fig. 5.9, the displacement is plotted for different values of $\alpha^D = \alpha^H = \alpha$ with constant $p^D = p^H$ and $q^D = q^H$. The highest damping is observed for $\alpha = 1$. The fractional orders

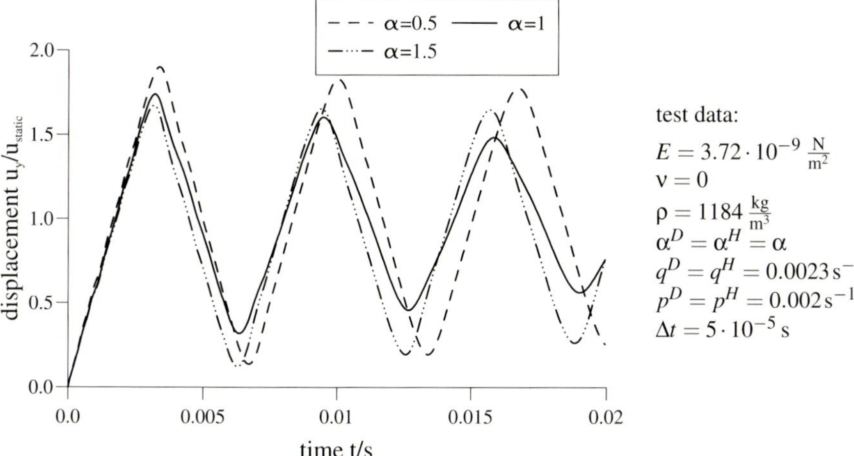

Fig. 5.9. Longitudinal displacement at the free end versus time: Influence of fractional derivative α

reduce the damping effects but change the wave velocities, $\alpha = 1.5$ increase and $\alpha = 0.5$ decrease the velocity. The initial value of the wave velocity, however, is not changed, which is in accordance with the definition (B.5).

In the final experiment, different viscoelastic behavior due to different stress-strain relations is studied. First, the extreme, but not realistic case of pure hydrostatic damping, i.e., the deviatoric part of the stress-strain relation is assumed to be elastic, is compared to the more realistic cases of equal damping mechanisms and the pure deviatoric case, i.e., the hydrostatic part of the stress-strain relation is assumed to be elastic. In Fig. 5.10, the displacement at point P is plotted versus time for these three cases. This experiment shows that the deviatoric part of the stress-strain relation has more influence on the damping behavior than the hydrostatic part. Obviously, if both parts involve damping the highest damping is achieved. The purely hydrostatic damping is close to the elastic case, but all damping mechanism stiffen the material, which is concluded from the observed higher wave velocities compared to the elastic ones.

5.4.2 Elastic concrete slab foundation on viscoelastic clay half space

The propagation of waves in an elastic concrete foundation slab ($1\,m \times 1\,m \times 0.5\,m$) bonded on a viscoelastic clay half space will be analyzed. Both domains are coupled

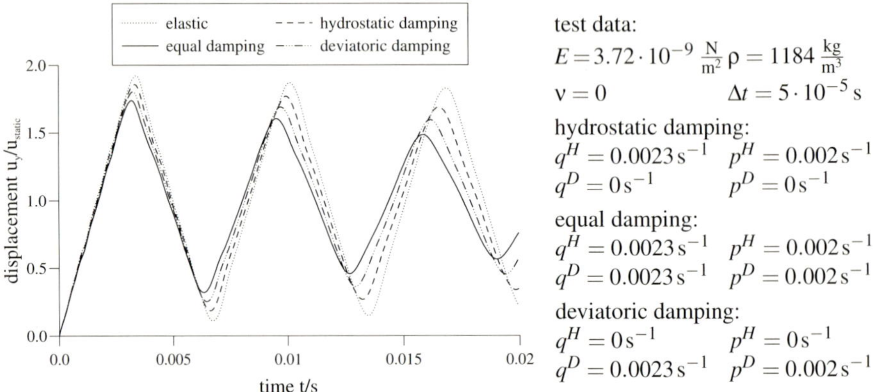

Fig. 5.10. Longitudinal displacement at the free end versus time: Comparison of modeling damping

by a substructure technique based on displacement- and traction-continuity at the interface. With this assumption, any uplifting or other nonlinear contact effects are neglected.

The problem geometry and the associated boundary discretization are shown in Fig. 5.11. The half space is discretized 2.5 m around the origin, which is enough

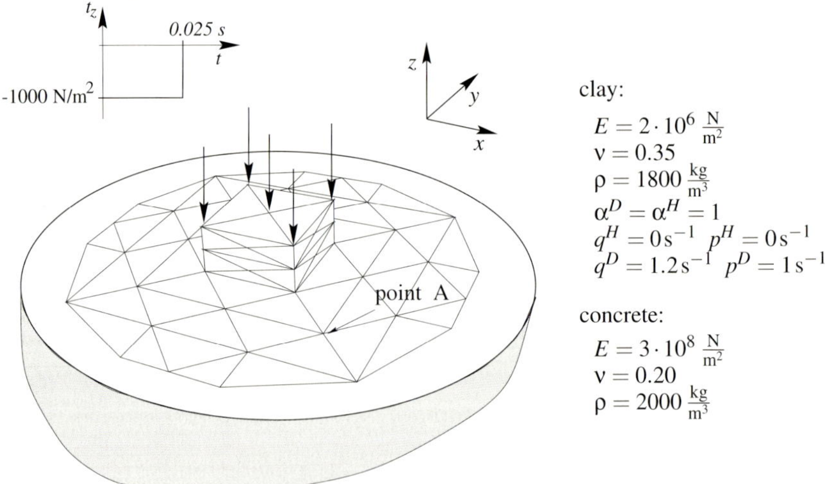

Fig. 5.11. Elastic concrete slab on viscoelastic clay half space: Boundary element discretization, material data, and loading function

to avoid truncation effects as shown in [76]. The top of the foundation slab is excited vertically by a pressure jump according to $t_z = -1000\,\mathrm{N/m}^2\,(H(t) - H(t - 0.025\,\mathrm{s}))$.

The remaining free surfaces of the slab and of the half space are traction free. The viscoelastic half space is modeled with the material data of clay, where only for the deviatoric part of the stress-strain relation damping is assumed. The spatial discretization is done with linear shape functions on triangles, and a BDF 2 is used as the underlying multistep method.

Due to the very different wave velocities of the two domains, in clay $c_1 = 42\,\mathrm{m/s}$, $c_2 = 20\,\mathrm{m/s}$ and in concrete $c_1 = 408\,\mathrm{m/s}$, $c_2 = 250\,\mathrm{m/s}$ and varying element sizes, the value β can not be chosen optimal for both domains. For a time step size of $\Delta t = 0.0025\,\mathrm{s}$ the value of β vary from $\beta = 0.15$ in the half space up to $\beta = 2$ in the foundation. In Fig. 5.12, the displacement at point A versus time is plotted for different time step sizes Δt. Slight oscillations are observed for the smallest

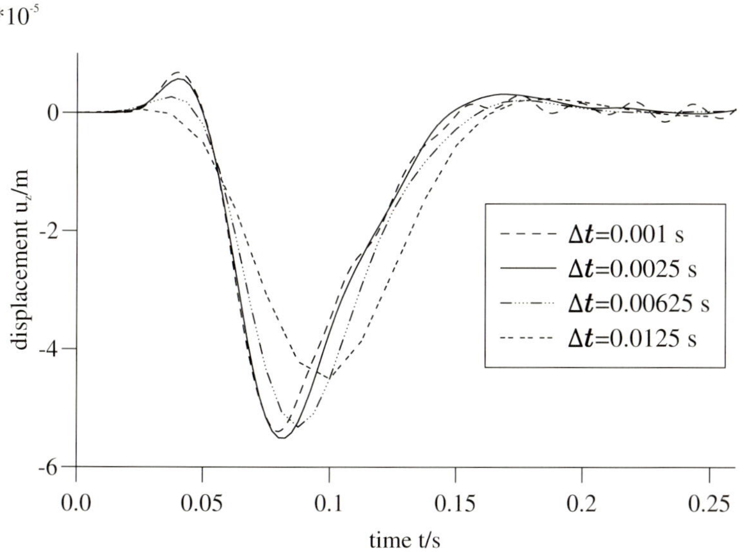

Fig. 5.12. Displacement u_z at point A versus time: Influence of time step size Δt

time step size $\Delta t = 0.001\,\mathrm{s}$ for increasing times t. However, the instabilities are far from those in the rod example. Moreover, if Δt is further reduced the oscillations do not increase strongly. This is caused by the damping influence of the infinite domain. In contrast, large values of Δt result in high numerical damping. But, this very coarse time approximation with $\Delta t = 0.0125\,\mathrm{s}$ corresponds to β values varying from $\beta_{min} = 0.75$ up to $\beta_{max} = 10$ which can not lead to satisfactory results. However, numerical convergence is observed for $\Delta t = 0.005\,\mathrm{s}$ or smaller. The shown graph for $\Delta t = 0.00625\,\mathrm{s}$ indicates the upper limit of reliable time step sizes.

In the next Fig. 5.13, the vertical displacement u_z at point A versus time is depicted for different viscous damping values q. As in the previous example, the great influence of the viscosity is observed. Increasing viscosity, i.e., larger values of q, stiffens the material, indicated by lower displacement amplitudes and higher

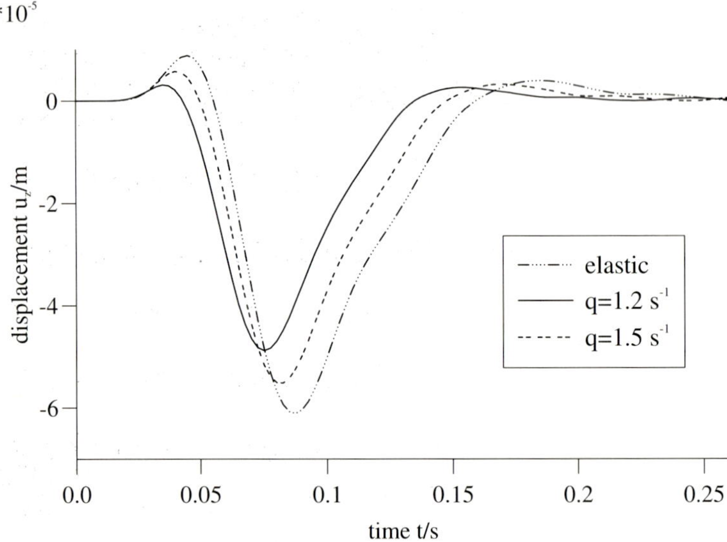

Fig. 5.13. Displacement u_z at point A versus time: Influence of damping values q

wave velocities. The results presented in Fig. 5.13 are calculated with time step size $\Delta t = 0.0025\,\mathrm{s}$ for all values of q. Varying q from 1 to 1.5 changes the compression wave velocity form $c_1 = 42\,\mathrm{m/s}$ to $c_1 = 45\,\mathrm{m/s}$ which makes an adapting of the time step size needless. The influence of the fractional exponent α is in this example very small contrary to the perspex rod. Reasons are found in the material clay as well as in the geometrical damping of the infinite domain and/or in the coupling with the elastic concrete slab.

The final parameter study concerns the modeling of the viscoelastic behavior. Above, it was assumed that only the deviatoric part of the stress-strain relation is viscoelastic, i.e., previously denoted as deviatoric damping. In Fig. 5.14, all three cases – hydrostatic damping, equal damping, and deviatoric damping – are compared. Obviously, pure hydrostatic damping has no influence on the displacement solution observed by the fact that an elastic modeling of the half space leads to the same results as the pure hydrostatic modeling. Consequently, the deviatoric case and damping in both parts of the stress-strain relation result in the same response function, i.e., only in the deviatoric part of the stress-strain relation the viscoelasticity has any influence on the results. These results allow the conclusion that damping caused by the material affects only the shear stress in a half space but not the normal – volumetric – stress. The hydrostatic – volumetric – stress is more affected by the geometrical damping. These conclusions are confirmed by a calculation of the same constellation – foundation slab on half space – where the half space was modeled by soil data.

Due to the lack of an analytical solution, next, the results of the proposed formulation are compared with another numerical method, the BEM using a calculation in

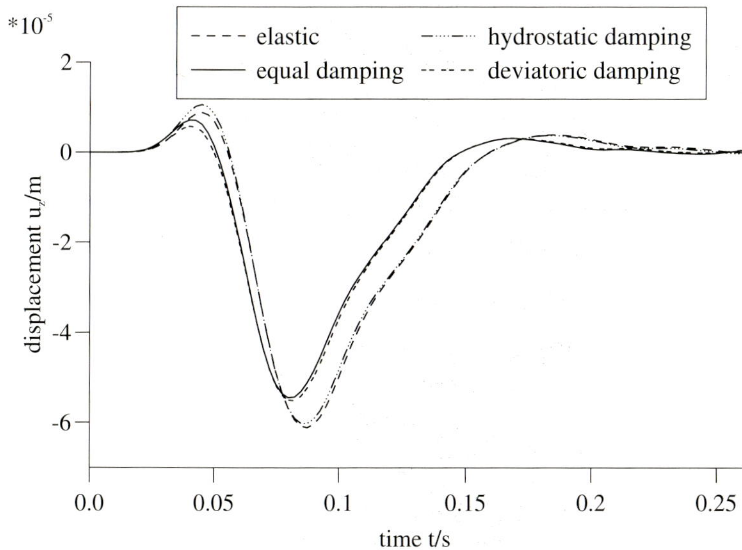

Fig. 5.14. Displacement u_z at point A versus time: Comparison of modeling damping

Laplace domain with subsequent inverse transformation. From the plenty of inverse Laplace transforms the method of Durbin [79] was chosen because several studies [133, 93, 53] indicate that this method is suitable for wave propagation problems. In Durbin's method, as in other inverse transformations, the parameters have to be chosen correctly, i.e., the real part of the Laplace variable α and a time period T must be determined. In the original paper [79], the values $\alpha T = 5$ and $T = 0.8 \cdot t_{max}$ which gives $\alpha = 20$ and $T = 0.8 \cdot 0.3\,\text{s} = 0.24\,\text{s}$ are recommended. Contrary to this, a series of numerical experiments here leads to $\alpha = 2$.

In Fig. 5.15, the vertical displacement at point A calculated using the proposed formulation and the inverse transformation for both parameters α is depicted versus time. For $t > 0.24\,\text{s}$ nothing interesting is observed, therefore, the time axis in Fig. 5.15 is truncated after $t = 0.24\,\text{s}$, whereas the calculation was performed until $t = 0.3\,\text{s}$. The zoom shows the first few seconds when the compression wave arrives at point A. There, clearly, the disadvantage of the inverse transformation is observed. Using $\alpha = 2$ gives a stable result for large times but a non causal behavior at small times. The compression wave should not arrive at point A before $t = 0.026\,\text{s}$. Contrary, $\alpha = 20$ reproduces a causal arrival of the wave front but leads to unstable results for large times, whereas the proposed method yields sufficient results for the complete observation period.

An interesting aspect in comparisons of numerical methods is the needed CPU-time. Because one time step needs the same time to calculate as one frequency of the Laplace formulation and these are the significant operations in the complete formulations, the efficiency is determined by the amount of used time steps or frequencies. To calculate the results in Fig. 5.15 either 150 time steps or 100 frequencies are

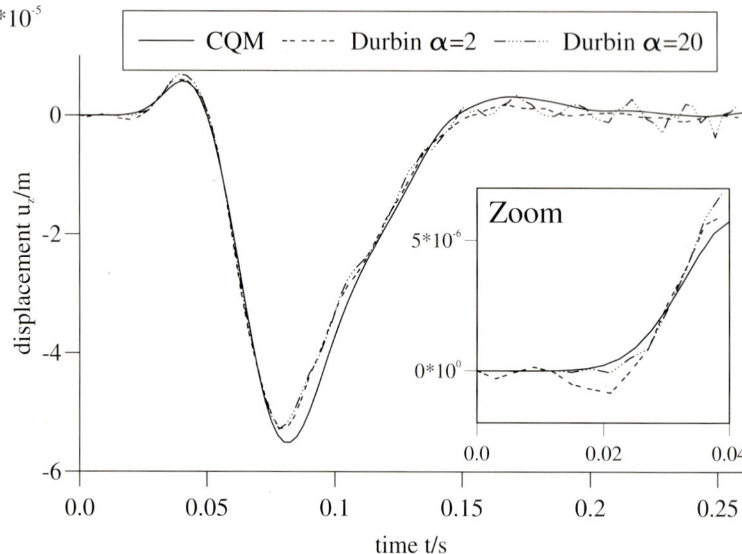

Fig. 5.15. Displacement u_z at point A versus time: Comparison of the proposed method (CQM) with a BEM formulation in Laplace domain with subsequent inverse transformation

used. From this the Laplace formulation seems to be faster but for the same good time resolution 150 frequencies would be necessary. As the amount of necessary frequencies depend also on the problem a general statement which formulation is more effective can not be given.

This problem shows that the proposed method is applicable even to more complex problems as a rod. Its insensitivity to the optimal choice of the time step size makes this method feasible to realistic applications where heavily differing element sizes or material data are possible.

6. Poroelastodynamic boundary element formulation

For a wide range of fluid infiltrated materials, such as water saturated soils, oil impregnated rocks, or air filled foams, the elastic theory is a crude approximation for investigating wave propagation in such media. The presence of a freely moving fluid in such materials modifies their mechanical response. Three mechanisms play a key role in the interaction between the interstitial fluid and the porous material:

- an increase of pore pressure induces a dilatation of the solid,
- a compression of the solid causes rise of pore pressure, if the fluid is prevented from escaping the pore network, and
- the relative motion between the solid and the fluid introduces energy dissipation to the otherwise conservative system.

These coupled mechanisms yield an apparent time-dependent character to the mechanical properties of the material, and, therefore, a different theory is necessary. A theory of porous materials containing a viscous fluid was presented by Biot [28]. This has been generally attributed as the starting point of the Theory of Poroelasticity. In the following years, Biot extended his theory to anisotropic case [29] and also to poroviscoelasticity [30]. The dynamic extension was done in two papers, one for low frequency range [31] and the other for high frequency range [32]. Among the significant findings was the identification of three waves for a three dimensional continuum, two compressional waves and one shear wave. This extra compressional wave, known as the slow wave, has been experimentally confirmed [144]. In Biot's theory a fully saturated material is assumed. The extension to a nearly saturated poroelastic solid was presented in [179].

A different approach to describe the dynamic behavior of porous media, known as the Theory of Porous Media [80], is based on the theory of mixtures and derived from the well known methods of continuum mechanics. It has been demonstrated that under small deformations, and some other restrictions, this and Biot's theory lead to the same governing equations [83]. Although Biot's theory is more based on physical intuition, it has the widest acceptance in geophysics and geomechanics.

6.1 Biot's theory of poroelasticity

6.1.1 Elastic skeleton

Following Biot's approach to model the behavior of porous media, an elastic skeleton with a statistical distribution of interconnected pores is considered [29]. This porosity is denoted by

$$\phi = \frac{V^f}{V}, \tag{6.1}$$

where V^f is the volume of the interconnected pores contained in a sample of bulk volume V. Contrary to these pores the sealed pores will be considered as part of the solid. As mentioned above, full saturation is assumed leading to $V = V^f + V^s$ with V^s the volume of the solid, i.e., a two-phase material is given.

If the constitutive equations are formulated for the elastic solid and the viscous interstitial fluid, a partial stress formulation is obtained [29]

$$\sigma_{ij}^s = 2G\varepsilon_{ij}^s + \left(K - \frac{2}{3}G + \frac{Q^2}{R}\right)\varepsilon_{kk}^s\delta_{ij} + Q\varepsilon_{kk}^f\delta_{ij} \tag{6.2a}$$

$$\sigma^f = -\phi p = Q\varepsilon_{kk}^s + R\varepsilon_{kk}^f, \tag{6.2b}$$

with $()^s$ and $()^f$ indicating either solid or fluid, respectively. The elastic skeleton is assumed to be isotropic and homogeneous with the two material constants compression K and shear modulus G known from elasticity. The coupling between the solid and the fluid is characterized by the two parameters Q and R. In the above, the sign conventions for stress and strain follow that of elasticity, namely, tensile stress and strain is denoted positive. Therefore, in equation (6.2b) the pore pressure p is the negative hydrostatic stress in the fluid σ^f.

An alternative representation of the constitutive equation (6.2) is used in Biot's earlier work [28]. There, the total stress $\sigma_{ij} = \sigma_{ij}^s + \sigma^f$ is introduced and with Biot's effective stress coefficient $\alpha = \phi(1 + Q/R)$ the constitutive equation with the solid strain $\varepsilon_{ij}^s = \varepsilon_{ij}$ and the pore pressure p

$$\sigma_{ij} = 2G\varepsilon_{ij} + \left(K - \frac{2}{3}G\right)\varepsilon_{kk}\delta_{ij} - \alpha\delta_{ij}p \tag{6.3}$$

is obtained (For simplicity, now, and in the following, the index s at the strain is skipped). Additional to the total stress σ_{ij}, as a second constitutive equation the variation of fluid volume per unit reference volume ζ is introduced

$$\zeta = \alpha\varepsilon_{kk} + \frac{\phi^2}{R}p. \tag{6.4}$$

This variation of fluid ζ is defined by the mass balance over a reference volume, i.e., by the continuity equation

$$\frac{\partial \zeta}{\partial t} + q_{i,i} = a \tag{6.5a}$$

or in the time integrated form

$$\zeta + \phi v_{i,i} = \int_0^t a(\tau)\,\mathrm{d}\tau \tag{6.5b}$$

with the specific flux $q_i = \phi \partial v_i / \partial t$, the relative fluid to solid displacement v_i, and a source term $a(t)$. Equation (6.5b) identify ζ as a kind of strain describing the motion of the fluid relative to the solid which takes a source in the fluid into account.

Additional to the fluid balance (6.5a), the balance of momentum for the bulk material must be fulfilled. This dynamic equilibrium is given by

$$\sigma_{ij,j} + F_i = \rho \frac{\partial^2 u_i}{\partial t^2} + \phi \rho_f \frac{\partial^2 v_i}{\partial t^2}\,, \tag{6.6}$$

with the bulk body force per unit volume F_i, the solid displacements u_i, and the bulk density $\rho = \rho_s (1 - \phi) + \phi \rho_f$. The density of the solid and the fluid is denoted by ρ_s and ρ_f, respectively. The relation of the solid strain to the solid displacement is chosen linear

$$\varepsilon_{ij} = \frac{1}{2}\left(u_{i,j} + u_{j,i}\right) \tag{6.7}$$

assuming small deformation gradients.

Next, the fluid transport in the interstitial space expressed by the specific flux q_i is modeled with a generalized Darcy's law

$$q_i = -\kappa \left(p_{,i} + \rho_f \frac{\partial^2 u_i}{\partial t^2} + \frac{\rho_a + \phi \rho_f}{\phi} \frac{\partial^2 v_i}{\partial t^2} \right) \tag{6.8}$$

where κ denotes the permeability. In equation (6.8), an additional density the apparent mass density ρ_a is introduced by Biot [31] to describe the interaction between fluid and skeleton. It can be written as $\rho_a = C \phi \rho_f$ where C is a factor depending on the geometry of the pores and the frequency of excitation. At low frequency, Bonnet and Auriault [36] measured $C = 0.66$ for a sphere assembly of glass bead. In higher frequency ranges, a certain functional dependence of C on frequency has been proposed based on conceptual porosity structures, e.g., in [32] and [36]. In the following, if nothing different is noted, $C = 0.66$ is assumed.

Aiming at the equation of motion, the above balance laws and constitutive equations have to be combined. To do this, first, the degrees of freedom must be determined. Here, there are several possibilities: i) to use the solid displacements u_i and the relative fluid to solid displacement v_i (six unknowns in 3-d) or ii) a combination of the pore pressure p and the solid displacements u_i (four unknowns in 3-d). As shown in [35], it is sufficient to use the latter choice, i.e., the solid displacements u_i

and the pore pressure p will become basic variables to describe a poroelastic continuum. Therefore, the above equations are reduced to these four unknowns. First, Darcy's law (6.8) is rearranged to obtain v_i. Since v_i is given as second time derivative in (6.8), this is only possible in Laplace domain. After transformation to Laplace domain, the relative fluid to solid displacement is

$$\hat{v}_i = -\underbrace{\frac{\kappa\rho_f\phi^2 s^2}{\phi^2 s + s^2\kappa(\rho_a + \phi\rho_f)}}_{\beta} \frac{1}{s^2\phi\rho_f} \left(\hat{p}_{,i} + s^2\rho_f\hat{u}_i\right) . \tag{6.9}$$

In equation (6.9), the abbreviation β is defined for further usage. Moreover, vanishing initial conditions for u_i and v_i are assumed here and in the following. Now, the final set of differential equations for the displacement $\hat{u}_i$ and the pore pressure $\hat{p}$ is obtained by inserting the constitutive equations (6.3) and (6.4) in the Laplace transformed dynamic equilibrium (6.6) and continuity equation (6.5a) with $\hat{v}_i$ from equation (6.9). This leads to the final set of differential equations for the displacement $\hat{u}_i$ and the pore pressure $\hat{p}$

$$G\hat{u}_{i,jj} + \left(K + \frac{1}{3}G\right)\hat{u}_{j,ij} - (\alpha - \beta)\,\hat{p}_{,i} - s^2(\rho - \beta\rho_f)\,\hat{u}_i = -\hat{F}_i \tag{6.10}$$

$$\frac{\beta}{s\rho_f}\hat{p}_{,ii} - \frac{\phi^2 s}{R}\hat{p} - (\alpha - \beta)\, s\hat{u}_{i,i} = -\hat{a} . \tag{6.11}$$

This set of equations describes the behavior of a poroelastic continuum completely. However, an analytical representation in time domain is only possible for $\kappa \to \infty$. This case would represent a negligible friction between solid and interstitial fluid.

6.1.2 Viscoelastic skeleton

In the constitutive equations above (6.3) and (6.4), the only damping effects taken into account are caused by the interaction of the viscous fluid and the elastic solid. Introducing additionally viscoelasticity is done by means of the elastic-viscoelastic correspondence principle, as shown by Biot [30]. In a typical implementation in Laplace domain, the material constants shown in (6.3) and (6.4) are replaced by the corresponding functions of the Laplace variable. However, this approach provides little physical insight into the rheological models introduced, because the effective stress coefficient α and R or the pair of coefficients Q and R in the partial stress formulation, have no simple relation to the compression or shear behavior of the constituents. Rather, considerations of constitutive relation at micro mechanical level [69] lead to a more rational model for our purpose

$$\alpha = 1 - \frac{K}{K_s} \quad \text{and} \tag{6.12a}$$

$$R = \frac{\phi^2 K_f K_s^2}{K_f(K_s - K) + \phi K_s(K_s - K_f)} , \tag{6.12b}$$

where K_s denotes the compression modulus of the solid grains and K_f the compression modulus of the fluid. With these expressions it is possible to discuss how to implement viscoelastic behavior from a physical point of view.

In the explicit form of (6.12), a viscoelastic model can be applied to each of the moduli, corresponding to different physical effects. In detail:

- Replacing G by the complex modulus $\hat{G}(s)$ models a viscoelastic shear behavior of the solid frame.
- Replacing K_s by the complex modulus $\hat{K}_s(s)$ models a viscoelastic behavior of solid grains against volumetric deformation. This is necessary if the material has its own damping mechanism.
- Replacing K by the complex modulus $\hat{K}(s)$ models a viscoelastic behavior of the solid skeleton against volumetric deformation. Such a behavior can be caused, e.g., by micro-pores which are not connected to the main part of the fluid. The fluid in micro-pores can propagate through micro-cracks in the material causing damping due to the time required to reach localized equilibrium.
- Replacing K_f by the complex modulus $\hat{K_f}(s)$ models a viscoelastic behavior of the fluid. This however will not be attempted here for the following reasons: First, most pore fluids such as water or air are not viscoelastic. Second, a viscoelastic fluid can have shear stresses, which will interact with the surrounding solid. These effects are not modeled in Biot's theory. An arbitrarily generalization will not lead to a consistent theory.

Summarizing, in the following, a time-dependent compression and shear modulus of the solid $\hat{K}_s(s)$ and $\hat{G}(s)$ and a time-dependent bulk modulus $\hat{K}(s)$ are taken into account. This leads to the poroviscoelastic constitutive equations in Laplace domain as

$$\hat{\sigma}_{ij} = 2\hat{G}\hat{\varepsilon}_{ij} + \left(\hat{K} - \frac{2}{3}\hat{G}\right)\hat{\varepsilon}_{kk}\delta_{ij} - \hat{\alpha}\delta_{ij}\hat{p} \tag{6.13}$$

$$\hat{\zeta} = \hat{\alpha}\hat{\varepsilon}_{kk} + \frac{\phi^2}{\hat{R}}\hat{p}\,, \tag{6.14}$$

with

$$\hat{\alpha}(s) = 1 - \frac{\hat{K}(s)}{\hat{K}_s(s)} \qquad \text{and} \tag{6.15a}$$

$$\hat{R}(s) = \frac{\phi^2 K_f \hat{K}_s^2(s)}{K_f\left(\hat{K}_s(s) - \hat{K}(s)\right) + \phi\hat{K}_s(s)\left(\hat{K}_s(s) - K_f\right)}\,. \tag{6.15b}$$

Note, every formerly constant which is now indicated with $\hat{()}$ is a function of s, respectively of time. In the following, it is assumed that $\hat{K}_s(s)$, $\hat{G}(s)$, and $\hat{K}(s)$ are modeled as a three-parameter model (see Fig. 5.1) using the correspondence relation (5.17)

$$\hat{K}(s) = K\frac{1+q_k s^{\alpha^k}}{1+p_k s^{\alpha^k}} \quad \hat{K}_s(s) = K_s\frac{1+q_{ks} s^{\alpha^{ks}}}{1+p_{ks} s^{\alpha^{ks}}} \quad \hat{G}(s) = G\frac{1+q_g s^{\alpha^g}}{1+p_g s^{\alpha^g}}\,. \tag{6.16}$$

This completes the constitutive equations for a poroviscoelastic model.

Now, to achieve the governing differential equations for a poroviscoelastic continuum the elastic moduli in equations (6.10) and (6.11) are replaced by the corresponding complex moduli of equation (6.16), respectively. This leads to the final set of differential equations for the displacement $\hat{u}_i$ and the pore pressure $\hat{p}$

$$\hat{G}\hat{u}_{i,jj} + \left(\hat{K} + \frac{1}{3}\hat{G}\right)\hat{u}_{j,ij} - (\hat{\alpha} - \beta)\,\hat{p}_{,i} - s^2\,(\rho - \beta\rho_f)\,\hat{u}_i = -\hat{F} \tag{6.17}$$

$$\frac{\beta}{s\rho_f}\hat{p}_{,ii} - \frac{\phi^2 s}{\hat{R}}\hat{p} - (\hat{\alpha} - \beta)\, s\hat{u}_{i,i} = -\hat{a}\,. \tag{6.18}$$

With this set of equations the dynamic behavior of a poroviscoelastic continuum is completely defined. (The functional argument (s) is dropped in (6.17) and (6.18) for brevity.)

6.2 Fundamental solutions

In the following, for a poroelastodynamic boundary element formulation fundamental solutions are necessary, especially, to treat wave propagation problems time-dependent solutions. In general, these are not available in closed form, except for the special case of an inviscid fluid, $\kappa \to \infty$ [184]. Fortunately, a time-dependent boundary element formulation based on the convolution quadrature method needs only Laplace transformed fundamental solutions. These are presented in the literature for 2-d [45] and 3-d [46] using solid displacements and pore pressure as unknowns. For the other case, solid and fluid displacements as unknowns, the fundamental solutions can be found in [124].

However, there is some possible misunderstanding about the relation which fundamental solution correspond to which integral formulation, due to the fact that the poroelastodynamic operator is not self-adjoint. It is necessary to recall the derivation of fundamental solutions to gain confidence and resolve inconsistencies in some papers as reported in [74]. In principle, two possibilities exist: i) using the analogy between thermo- and poroelasticity in Laplace or Fourier domain to convert the thermoelastic solutions to poroelastic ones [75], or ii) the method of Hörmander [105]. The latter will be used here.

Central idea in the method of Hörmander. The method of Hörmander can be applied to every elliptic set of coupled differential equations with constant coefficients. Such systems can be written in short form

$$\mathbf{B}\mathbf{u} = \mathbf{0} \tag{6.19}$$

with the matrix differential operator $\mathbf{B}$ and the vector of unknowns $\mathbf{u}$. A matrix differential operator is a matrix with elements of differential operators. A multiplication of such a matrix with a vector or a matrix represents that the elements, differential operators, are applied on the elements of the vector or matrix following the rules of a normal matrix with vector or matrix multiplication, e.g.,

$$\mathbf{Bu} = \begin{pmatrix} \partial_x & \partial_y & 0 \\ \partial_t & 4 & \partial_y \\ 0 & \partial_x & 2\partial_y \end{pmatrix} \begin{pmatrix} u(x,y,t) \\ v(x,y,t) \\ w(x,y,t) \end{pmatrix} \Leftrightarrow \begin{matrix} \dfrac{\partial u(x,y,t)}{\partial x} + \dfrac{\partial v(x,y,t)}{\partial y} \\ \dfrac{\partial u(x,y,t)}{\partial t} + 4v(x,y,t) + \dfrac{\partial w(x,y,t)}{\partial y} \\ \dfrac{\partial v(x,y,t)}{\partial x} + 2\dfrac{\partial w(x,y,t)}{\partial y} \end{matrix} .$$

Rules known from the matrix calculus can be transformed analogous to matrix differential operator. In the following, the matrix of *cofactors* $\mathbf{B}^{co}$ is used to calculate the inverse matrix to $\mathbf{B}$

$$\mathbf{BB}^{-1} = \mathbf{I} \quad \text{with} \quad \mathbf{B}^{-1} = \frac{\mathbf{B}^{co}}{\det(\mathbf{B})} . \tag{6.20}$$

$\mathbf{B}^{co}$ is sometimes denoted the adjoint matrix and their elements are calculated following the rule: The element B^{co}_{ij} is the determinant of the matrix $\mathbf{B}$ without the column j and row i multiplied by $(-1)^{i+j}$ (see, e.g., [13]). In the above example, the first two elements are

$$B^{co}_{11} = (-1)^2 \begin{vmatrix} 4 & \partial_y \\ \partial_x & 2\partial_y \end{vmatrix} = 8\partial_y - \partial_{xy} \qquad B^{co}_{21} = (-1)^3 \begin{vmatrix} \partial_t & \partial_y \\ 0 & 2\partial_y \end{vmatrix} = -2\partial_t\partial_y$$

and the complete matrix of cofactors

$$\mathbf{B}^{co} = \begin{pmatrix} 8\partial_y - \partial_{xy} & -2\partial_{yy} & \partial_{yy} \\ -2\partial_t\partial_y & 2\partial_{xy} & -\partial_{xy} \\ \partial_t\partial_x & -2\partial_{xx} & 4\partial_x - \partial_t\partial_y \end{pmatrix} .$$

Note, this matrix is different to the adjoint operator.

For the poroelastodynamic governing equations (6.10) and (6.11) the matrix differential operator $\mathbf{B}$ is

$$\mathbf{B} = \begin{bmatrix} G\nabla^2 + \left(K + \frac{1}{3}G\right)\partial_i\partial_j - s^2(\rho - \beta\rho_f) & -(\alpha - \beta)\partial_i \\ -s(\alpha - \beta)\partial_j & \dfrac{\beta}{s\rho_f}\nabla^2 - \dfrac{\phi^2 s}{R} \end{bmatrix} \quad \text{and} \quad \mathbf{u} = \begin{bmatrix} u_i \\ p \end{bmatrix} . \tag{6.21}$$

In equation (6.21), ∂_i denotes the partial derivative with respect to x_i and $\nabla^2 = \partial_i\partial_i$ is the Laplace operator. The elements on the secondary diagonal indicate that this operator is not self-adjoint, which is caused by the first time derivative represented by the multiplicative factor s. Physically, this represents dissipation due to the friction between solid and interstitial fluid leading to a loss of energy.

To find the fundamental solutions $\mathbf{G}$, equation (6.19) has to be solved with a Dirac distribution as inhomogeneity for every degree of freedom, i.e.,

$$\mathbf{B}\mathbf{G}+\mathbf{I}\delta(\mathbf{x}-\mathbf{y})=\mathbf{0} \tag{6.22}$$

has to be fulfilled (see definition B.1.1). Now, taking the ansatz

$$\mathbf{G}=\mathbf{B}^{co}\varphi \tag{6.23}$$

for the matrix of fundamental solutions with a unknown scalar function φ equation (6.22) can be rewritten

$$\begin{aligned}\mathbf{B}\mathbf{B}^{co}\varphi+\mathbf{I}\delta(\mathbf{x}-\mathbf{y}) &= \det(\mathbf{B})\,\mathbf{I}\varphi+\mathbf{I}\delta(\mathbf{x}-\mathbf{y})=\mathbf{0}\\ &\rightsquigarrow \det(\mathbf{B})\,\varphi+\delta(\mathbf{x}-\mathbf{y})=0\,.\end{aligned} \tag{6.24}$$

In equation (6.24), the definition (6.20) of an inverse matrix to $\mathbf{B}$ is used.

With the result (6.24), finding the fundamental solutions is reduced to find the scalar function φ. After determination of φ by backward substitution in equation (6.23) the matrix of the fundamental solutions is determined.

Fundamental solutions for poroelastodynamics. Now, after explaining the central idea of Hörmander's method the fundamental solutions of poroelastodynamics can be deduced. From the mathematical theory of Green's formula it is known that the fundamental solutions should satisfy the adjoint operator [173]. Opposite to elasticity the governing operator in poroelasticity is not self-adjoint. Therefore, here, the solution for the adjoint operator $\mathbf{B}^*$

$$\mathbf{B}^*\mathbf{G}+\mathbf{I}\delta(\mathbf{x}-\mathbf{y})=\mathbf{0} \tag{6.25}$$

is required with

$$\mathbf{B}^*=\begin{bmatrix}A+B\partial_{11} & B\partial_{12} & B\partial_{13} & sC\partial_1\\ B\partial_{12} & A+B\partial_{22} & B\partial_{23} & sC\partial_2\\ B\partial_{13} & B\partial_{23} & A+B\partial_{33} & sC\partial_3\\ C\partial_1 & C\partial_2 & C\partial_3 & D\end{bmatrix}\qquad \begin{aligned}A &= G\nabla^2-s^2(\rho-\beta\rho_f)\\ B &= K+\tfrac{1}{3}G\\ C &= \alpha-\beta\\ D &= \frac{\beta}{s\rho_f}\nabla^2-\frac{\phi^2 s}{R}\,.\end{aligned} \tag{6.26}$$

From the preceding paragraph it is obvious that first the unknown function φ has to be determined. For this, the determinant of the operator matrix $\mathbf{B}^*$ is calculated

$$\begin{aligned}\det(\mathbf{B}^*) &= A^2\left[\left(BD-C^2s\right)\nabla^2+AD\right]\\ &= \frac{\beta}{s\rho_f}\left(K+\frac{4}{3}G\right)G^2\left(\nabla^2-\frac{s^2(\rho-\beta\rho_f)}{G}\right)^2\\ &\left[\nabla^4-\left(\frac{\phi^2 s^2\rho_f}{\beta R}+\frac{s^2(\rho-\beta\rho_f)}{K+\frac{4}{3}G}+\frac{s^2\rho_f(\alpha-\beta)^2}{\beta\left(K+\frac{4}{3}G\right)}\right)\nabla^2+\frac{s^4\phi^2\rho_f(\rho-\beta\rho_f)}{\beta R\left(K+\frac{4}{3}G\right)}\right]\end{aligned} \tag{6.27}$$

This determinant has obviously three roots: the two resulting from the brackets above

$$\lambda_{1,2}^2 = \frac{1}{2}\left[\frac{\phi^2 s^2 \rho_f}{\beta R} + \frac{s^2(\rho-\beta\rho_f)}{K+\frac{4}{3}G} + \frac{s^2\rho_f(\alpha-\beta)^2}{\beta\left(K+\frac{4}{3}G\right)} \pm \sqrt{\left(\frac{\phi^2 s^2 \rho_f}{\beta R} + \frac{s^2(\rho-\beta\rho_f)}{K+\frac{4}{3}G} + \frac{s^2\rho_f(\alpha-\beta)^2}{\beta\left(K+\frac{4}{3}G\right)}\right)^2 - 4\frac{s^4\phi^2\rho_f(\rho-\beta\rho_f)}{\beta R\left(K+\frac{4}{3}G\right)}}\,\right] \tag{6.28a}$$

and the double root

$$\lambda_3^2 = \frac{s^2(\rho-\beta\rho_f)}{G} . \tag{6.28b}$$

These three roots correspond to the three expected waves – the fast and slow compressional wave to $\lambda_{1,2}$ and the shear wave to λ_3. Using them yields a representation of the determinant

$$\det(\mathbf{B}^*) = \frac{G^2\beta}{s\rho_f}\left(K+\frac{4}{3}G\right)\left(\nabla^2-\lambda_3\right)\left(\nabla^2-\lambda_3\right)\left(\nabla^2-\lambda_1\right)\left(\nabla^2-\lambda_2\right) . \tag{6.29}$$

This expression is inserted in equation (6.24) to determine φ. With the abbreviation $\psi = G^2\beta/(s\rho_f)(K+4/3G)\left(\nabla^2-\lambda_3\right)\varphi$ it becomes obvious that ψ and subsequently φ is found by solving a higher order Helmholtz equation

$$\psi\left(\nabla^2-\lambda_3\right)\left(\nabla^2-\lambda_1\right)\left(\nabla^2-\lambda_2\right)+\delta(\mathbf{x}-\mathbf{y}) = 0 . \tag{6.30}$$

The solution for either 2-d and 3-d is given in [54]. Here, only the 3-d solution

$$\psi = \frac{1}{4\pi r}\left[\frac{e^{-\lambda_1 r}}{\left(\lambda_1^2-\lambda_2^2\right)\left(\lambda_1^2-\lambda_3^2\right)} + \frac{e^{-\lambda_2 r}}{\left(\lambda_2^2-\lambda_1^2\right)\left(\lambda_2^2-\lambda_3^2\right)} + \frac{e^{-\lambda_3 r}}{\left(\lambda_3^2-\lambda_1^2\right)\left(\lambda_3^2-\lambda_2^2\right)}\right] \tag{6.31}$$

is presented.

In the next step, with equation (6.23) the fundamental solutions can be determined. To do so, the matrix of cofactors has to be calculated resulting in

$$\mathbf{B}^{*co} = A\begin{bmatrix} F(\partial_{22}+\partial_{33})+AD & -F\partial_{12} & -F\partial_{13} & -ACs\partial_1 \\ -F\partial_{12} & F(\partial_{11}+\partial_{33})+AD & -F\partial_{23} & -ACs\partial_2 \\ -F\partial_{13} & -F\partial_{23} & F(\partial_{11}+\partial_{22})+AD & -ACs\partial_3 \\ -AC\partial_1 & -AC\partial_2 & -AC\partial_3 & A\left(B\nabla^2+A\right) \end{bmatrix} \tag{6.32}$$

with $F = BD - C^2 s$. Now, all partial results are merged. The definition of ψ yields φ and subsequent applying the matrix of differential operators $\mathbf{B}^{*co}$ on φ as given in (6.23) the fundamental solutions of poroelastodynamics are obtained

$$\mathbf{G} = \begin{bmatrix} \hat{U}_{ij}^s & \hat{U}_i^f \\ \hat{P}_j^s & \hat{P}^f \end{bmatrix} = \frac{s\rho_f}{G\beta\left(K+\frac{4}{3}G\right)}\begin{bmatrix} \left(F\nabla^2+AD\right)\delta_{ij}-F\partial_{ij} & -ACs\partial_i \\ -AC\partial_i & A\left(B\nabla^2+A\right) \end{bmatrix}\psi . \tag{6.33}$$

The explicit expressions of the elements of **G** are listed in appendix B.1.2.

6.3 Poroelastic Boundary Integral Formulation

6.3.1 Boundary integral equation

The boundary integral equation for dynamic poroelasticity in Laplace domain can be obtained using either the corresponding reciprocal work theorem [48] or the weighted residuals formulation [75]. In the former chapters on elastic or viscoelastic boundary element formulations only one of both possibilities were presented. Here, in case of the not self-adjoint poroelastic operator both methods will be presented. They need different fundamental solutions, but naturally both methods result finally in the same integral equation.

Weighted residuals. The poroelastodynamic integral equation can be derived directly by equating the inner product of (6.10) and (6.11), written in matrix form with matrix **B** defined in (6.21), and the matrix of the fundamental solutions **G** to a null vector, i.e.,

$$\int_{\Omega} \mathbf{G}^T \mathbf{B} \begin{bmatrix} \hat{u}_i \\ \hat{p} \end{bmatrix} \mathrm{d}\Omega = \mathbf{0} \qquad \text{with} \qquad \mathbf{G} = \begin{bmatrix} \hat{U}^s_{ij} & \hat{U}^f_i \\ \hat{P}^s_j & \hat{P}^f \end{bmatrix}, \tag{6.34}$$

where the integration is performed over a domain Ω with boundary Γ and vanishing body forces F_i and sources a are assumed. By this inner product, essentially, the error in satisfying the governing differential equations (6.10) and (6.11) is forced to be orthogonal to **G**. According to the theory of Green's formula and using partial integration the operator **B** is transformed from acting on the vector of unknowns $[\hat{u}_i \; \hat{p}]^T$ to the matrix of fundamental solutions **G**. These steps are easier understood looking at equation (6.34) written in index notation. This results in three (two) integral equations for the solid ($j = 1,2,3$ in 3-d and $j = 1,2$ in 2-d)

$$\begin{aligned} \int_{\Omega} \Bigg[& G\hat{u}_{i,kk}\hat{U}^s_{ij} + \left(K + \frac{1}{3}G\right)\hat{u}_{k,ik}\hat{U}^s_{ij} - (\alpha - \beta)\hat{p}_{,i}\hat{U}^s_{ij} - s^2(\rho - \beta\rho_f)\hat{u}_i\hat{U}^s_{ij} \\ & + \frac{\beta}{s\rho_f}\hat{p}_{,kk}\hat{P}^s_j - \frac{\phi^2 s}{R}\hat{p}\hat{P}^s_j - (\alpha - \beta)s\hat{u}_{k,k}\hat{P}^s_j \Bigg] \mathrm{d}\Omega = 0 \end{aligned} \tag{6.35}$$

and one integral equation for the fluid

$$\begin{aligned} \int_{\Omega} \Bigg[& G\hat{u}_{i,kk}\hat{U}^f_i + \left(K + \frac{1}{3}G\right)\hat{u}_{k,ik}\hat{U}^f_i - (\alpha - \beta)\hat{p}_{,i}\hat{U}^f_i - s^2(\rho - \beta\rho_f)\hat{u}_i\hat{U}^f_i \\ & + \frac{\beta}{s\rho_f}\hat{p}_{,kk}\hat{P}^f - \frac{\phi^2 s}{R}\hat{p}\hat{P}^f - (\alpha - \beta)s\hat{u}_{k,k}\hat{P}^f \Bigg] \mathrm{d}\Omega = 0\,. \end{aligned} \tag{6.36}$$

In the above integral equations, either one or two differentiations have to be transformed by either one or two partial integrations. Two exemplary parts of integral equations (6.35) and (6.36) are presented in detail to show the principal procedure.

All other partial integrations for the other parts in integral equations (6.35) and (6.36) can be performed analogously.

First, an integral with one differentiation in the kernel leads to (n_k is the outward normal vector)

$$\int_\Omega (\alpha-\beta)\, s\hat{u}_{k,k}\hat{P}^f \mathrm{d}\Omega = \int_\Gamma (\alpha-\beta)\, s\hat{u}_k n_k \hat{P}^f \mathrm{d}\Gamma - \int_\Omega (\alpha-\beta)\, s\hat{u}_k \hat{P}^f_{,k} \mathrm{d}\Omega \tag{6.37}$$

while an integral with two differentiation is transformed to

$$\begin{aligned} \int_\Omega G\hat{u}_{i,kk}\hat{U}^s_{ij}\mathrm{d}\Omega &= \int_\Gamma G\hat{u}_{i,k}n_k\hat{U}^s_{ij}\mathrm{d}\Gamma - \int_\Omega G\hat{u}_{i,k}\hat{U}^s_{ij,k}\mathrm{d}\Omega \\ &= \int_\Gamma G\hat{u}_{i,k}n_k\hat{U}^s_{ij}\mathrm{d}\Gamma - \int_\Gamma G\hat{u}_i\hat{U}^s_{ij,k}n_k\mathrm{d}\Gamma + \int_\Omega G\hat{u}_i\hat{U}^s_{ij,kk}\mathrm{d}\Omega\,. \end{aligned} \tag{6.38}$$

In both integrations by parts the divergence theorem is used. Obviously, one integration by parts changes the sign of the resulting domain integral while it remains unchanged in the case of two integration by parts, i.e., the operator $\mathbf{B}$ is transformed into its adjoint operator $\mathbf{B}^*$. This yields the following system of integral equations given in matrix notation as

$$\int_\Gamma \begin{bmatrix} \hat{U}^s_{ij} & -\hat{P}^s_j \\ \hat{U}^f_i & -\hat{P}^f \end{bmatrix} \begin{bmatrix} \hat{t}_i \\ \hat{q} \end{bmatrix} \mathrm{d}\Gamma - \int_\Gamma \begin{bmatrix} \hat{T}^s_{ij} & \hat{Q}^s_j \\ \hat{T}^f_i & \hat{Q}^f \end{bmatrix} \begin{bmatrix} \hat{u}_i \\ \hat{p} \end{bmatrix} \mathrm{d}\Gamma = -\int_\Omega (\mathbf{B}^*\mathbf{G})^T \begin{bmatrix} \hat{u}_i \\ \hat{p} \end{bmatrix} \mathrm{d}\Omega = \begin{bmatrix} \hat{u}_j \\ \hat{p} \end{bmatrix}. \tag{6.39}$$

To solve the domain integral in equation (6.39) for $\mathbf{y} \in \Omega$, the definition of fundamental solutions (6.25) and the property of the Dirac distribution (A.6) is used. Additionally, the traction vector $\hat{t}_i = \hat{\sigma}_{ij}n_j$ and the flux $\hat{q} = -\beta/(s\rho_f)\left(\hat{p}_{,i} + \rho_f s^2 \hat{u}_i\right) n_i$ is introduced, and the abbreviations

$$\hat{T}^s_{ij} = \left[\left(\left(K - \frac{2}{3}G\right)\hat{U}^s_{kj,k} + \alpha s\hat{P}^s_j\right)\delta_{i\ell} + G\left(\hat{U}^s_{ij,\ell} + \hat{U}^s_{\ell j,i}\right)\right] n_\ell \tag{6.40a}$$

$$\hat{Q}^s_j = \frac{\beta}{s\rho_f}\left[\hat{P}^s_{j,i} - \rho_f s\hat{U}^s_{ji}\right] n_i \tag{6.40b}$$

$$\hat{T}^f_i = \left[\left(\left(K - \frac{2}{3}G\right)\hat{U}^f_{k,k} + \alpha s\hat{P}^f\right)\delta_{i\ell} + G\left(\hat{U}^f_{i,\ell} + \hat{U}^f_{\ell,i}\right)\right] n_\ell \tag{6.40c}$$

$$\hat{Q}^f = \frac{\beta}{s\rho_f}\left[\hat{P}^f_{,j} - \rho_f s\hat{U}^f_j\right] n_j \tag{6.40d}$$

are used, where (6.40a) and (6.40b) can be interpreted as being the adjoint term to the traction vector $\hat{t}_i$ and the flux $\hat{q}$, respectively. With the fundamental solutions calculated in Sect. 6.2 or the explicit form given in appendix B.1.2, the integral representation deduced starting from the weighted residuals is completely given.

Reciprocal work theorem. Two poroelastic states defined by the solid strain ε_{ij}, total stress σ_{ij}, pore pressure p, and the variation of fluid volume ζ are introduced, where, to distinguish both states, one is indicated by $()'$. The reciprocity relation in Laplace domain

$$\hat{\sigma}_{ij}\hat{\varepsilon}'_{ij} + \hat{p}\hat{\zeta}' = \hat{\sigma}'_{ij}\hat{\varepsilon}_{ij} + \hat{p}'\hat{\zeta} \tag{6.41}$$

is proven to be equivalent by

$$\begin{aligned} \hat{\sigma}_{ij}\hat{\varepsilon}'_{ij} + \hat{p}\hat{\zeta}' &= C_{ijkl}\hat{\varepsilon}_{kl}\hat{\varepsilon}'_{ij} - \alpha\delta_{ij}\hat{p}\hat{\varepsilon}'_{ij} + \hat{p}\left(\alpha\delta_{ij}\hat{\varepsilon}'_{ij} + \frac{\phi^2}{R}\hat{p}'\right) \\ &= C_{klij}\hat{\varepsilon}_{ij}\hat{\varepsilon}'_{kl} - \alpha\delta_{ij}\hat{p}'\hat{\varepsilon}_{ij} + \hat{p}'\alpha\delta_{ij}\hat{\varepsilon}_{ij} + \frac{\phi^2}{R}\hat{p}' = \hat{\sigma}'_{ij}\hat{\varepsilon}_{ij} + \hat{p}'\hat{\zeta} \end{aligned} \tag{6.42}$$

using the constitutive equation (6.4) for ζ' and symmetry of the elasticity tensor C_{ijkl}. Integration of (6.41) over the domain Ω yields the reciprocal work theorem of poroelasticity, a generalization of Betti's reciprocal work theorem for elasticity [52]

$$\int_\Omega \left(\hat{\sigma}_{ij}\hat{\varepsilon}'_{ij} + \hat{p}\hat{\zeta}'\right) \mathrm{d}\Omega = \int_\Omega \left(\hat{\sigma}'_{ij}\hat{\varepsilon}_{ij} + \hat{p}'\hat{\zeta}\right) \mathrm{d}\Omega\,. \tag{6.43}$$

An inverse transformation would lead to the time-dependent reciprocal work theorem where products in (6.43) would be convolutions in time. However, for deducing the boundary integral equation it is more suitable to stay in Laplace domain.

In the following, parts of the integral in equation (6.43) will be treated separately. The first part on the left hand side of equation (6.43) is transformed

$$\begin{aligned} \int_\Omega \hat{\sigma}_{ij}\hat{\varepsilon}'_{ij}\mathrm{d}\Omega = \int_\Omega \hat{\sigma}_{ij}\hat{u}'_{i,j}\mathrm{d}\Omega &= \int_\Gamma \hat{\sigma}_{ij}n_j\hat{u}'_i\mathrm{d}\Gamma - \int_\Omega \hat{\sigma}_{ij,j}\,\hat{u}'_i\mathrm{d}\Omega \\ &= \int_\Gamma \hat{t}_i\hat{u}'_i\mathrm{d}\Gamma - \int_\Omega \left(\rho s^2\hat{u}_i + \phi\rho_f s^2\hat{v}_i - \hat{F}_i\right)\hat{u}'_i\mathrm{d}\Omega\,. \end{aligned} \tag{6.44}$$

by considering the linear strain-displacement relation (6.7), the divergence theorem, and the dynamic equilibrium (6.6). The remaining integral on the left hand side of equation (6.43) can be rearranged using the constitutive equation (6.5a) and the divergence theorem

$$\int_\Omega \hat{p}\hat{\zeta}'\mathrm{d}\Omega = \int_\Omega \hat{p}\frac{-\hat{q}_{i,i} + \hat{a}}{s}\mathrm{d}\Omega = -\frac{1}{s}\int_\Gamma \hat{p}\hat{q}'\mathrm{d}\Gamma + \frac{1}{s}\int_\Omega \hat{p}_{,i}\hat{q}'_i\mathrm{d}\Omega + \int_\Omega \hat{p}\frac{\hat{a}'}{s}\mathrm{d}\Omega\,. \tag{6.45}$$

In the second integral on the right hand side of equation (6.45), Darcy's law for $\hat{p}_i$ and the flux $\hat{q}_i = \phi s\hat{v}_i$ is inserted

$$\frac{1}{s}\int_\Omega \hat{p}_{,i}\hat{q}'_i\mathrm{d}\Omega = -\int_\Omega \left(\frac{\hat{q}_i}{\kappa} + s^2\rho_f\hat{u}_i + s^2\frac{\rho_a + \phi\rho_f}{\phi}\hat{v}_i\right)\phi\hat{v}'_i\mathrm{d}\Omega\,. \tag{6.46}$$

These steps, (6.44) - (6.46), have as well to be performed on the right hand side of the reciprocal work theorem (6.43). Then, after gathering all intermediary results, several terms cancel each other and, finally, the integral equation

$$\int_\Gamma \left[\hat{t}_i \hat{u}_i' - \hat{t}_i' \hat{u}_i\right] \mathrm{d}\Gamma - \frac{1}{s} \int_\Gamma \left[\hat{p}\hat{q}' - \hat{p}'\hat{q}\right] \mathrm{d}\Gamma + \int_\Omega \left[\hat{F}_i \hat{u}_i' - \hat{F}_i' \hat{u}_i\right] \mathrm{d}\Omega + \frac{1}{s} \int_\Omega \left[\hat{p}\hat{a}' - \hat{p}'\hat{a}\right] \mathrm{d}\Omega = 0 \tag{6.47}$$

is obtained.

To find the solution, i.e., the three (two) displacement components $\hat{u}_i$ in 3-d (2-d) and the pore pressure $\hat{p}$, four (three) different loads, the forces $\hat{F}_{ij}'$ in the three (two) coordinate directions j and one source $\hat{a}'$, are applied to the primed state in integral equation (6.47). As usual in boundary integral formulations, these loads are point loadings, i.e., Dirac distributions, $\hat{F}_{ij}' = \delta(\mathbf{x}-\mathbf{y})\delta_{ij}$ and $\hat{a}' = \delta(\mathbf{x}-\mathbf{y})$. Assuming vanishing body forces and source terms of the unprimed state, the corresponding representation to (6.39) is achieved in a matrix notation as

$$\int_\Gamma \begin{bmatrix} \hat{u}_{ij}' & \frac{1}{s}\hat{p}_j' \\ -s\hat{u}_i' & -\hat{p}' \end{bmatrix} \begin{bmatrix} \hat{t}_i \\ \hat{q} \end{bmatrix} \mathrm{d}\Gamma - \int_\Gamma \begin{bmatrix} \hat{t}_{ij}' & \frac{1}{s}\hat{q}_j' \\ -s\hat{t}_i' & -\hat{q}' \end{bmatrix} \begin{bmatrix} \hat{u}_i \\ \hat{p} \end{bmatrix} \mathrm{d}\Gamma = \int_\Omega \begin{bmatrix} \hat{F}_{ij}' & 0 \\ 0 & \hat{a}' \end{bmatrix} \begin{bmatrix} \hat{u}_i \\ \hat{p} \end{bmatrix} \mathrm{d}\Omega = \begin{bmatrix} \hat{u}_j \\ \hat{p} \end{bmatrix} . \tag{6.48}$$

Since the loadings of the primed state are point loadings, the primed state represents the fundamental solutions of the governing differential equations (6.10) and (6.11) contrary to the weighted residual derivation, i.e., it has to fulfill the equation

$$\mathbf{B}\mathbf{G}' + \mathbf{I}\delta(\mathbf{x}-\mathbf{y}) = \mathbf{0} \quad \text{with} \quad \mathbf{G}' = \begin{bmatrix} \hat{u}_{ij}' & \hat{u}_i' \\ \hat{p}_j' & \hat{p}' \end{bmatrix} \tag{6.49}$$

with the original operator $\mathbf{B}$. Additionally, in integral equation (6.48) $\hat{t}_{ij}', \hat{t}_i'$ and $\hat{q}', \hat{q}_j'$ denotes the fundamental traction and flux solution, respectively

$$\hat{t}_{ij}' = \left[\left(\left(K - \frac{2}{3}G\right)\hat{u}_{kj,k}' - \alpha\hat{p}_j'\right)\delta_{i\ell} + G\left(\hat{u}_{ij,\ell}' + \hat{u}_{\ell j,i}'\right)\right] n_\ell \tag{6.50a}$$

$$\hat{q}_j' = -\frac{\beta}{s\rho_f}\left[\hat{p}_{j,i}' + \rho_f s^2 \hat{u}_{ji}'\right] n_i \tag{6.50b}$$

$$\hat{t}_i' = \left[\left(\left(K - \frac{2}{3}G\right)\hat{u}_{k,k}' - \alpha\hat{p}'\right)\delta_{i\ell} + G\left(\hat{u}_{i,\ell}' + \hat{u}_{\ell,i}'\right)\right] n_\ell \tag{6.50c}$$

$$\hat{q}' = -\frac{\beta}{s\rho_f}\left[\hat{p}_{,j}' + \rho_f s^2 \hat{u}_j'\right] n_j , \tag{6.50d}$$

which are different to the solutions (6.40). This is due to the fact that the fundamental solution $\mathbf{G}$ corresponds to the adjoint operator $\mathbf{B}^*$ and $\mathbf{G}'$ corresponds to the original operator **B**. Comparing the two equations (6.25) and (6.49) for the determination of **G** and **G**′, respectively, the relations

$$\hat{u}'_{ij} = \hat{U}^s_{ij} \quad -s\hat{u}'_i = \hat{U}^f_i \quad -\hat{p}'_j = s\hat{P}^s_j \quad \hat{p}' = \hat{P}^f \tag{6.51}$$

hold. With these relations, the equivalence of integral equation (6.39) deduced by the weighted residuals methodology and (6.49) deduced from the reciprocal work theorem is straight forward. In the following, the integral equation (6.39) with the fundamental solutions **G** (6.33) is taken to be the representation of choice.

Singular integral equation. When moving **y** to the boundary Γ to determine the unknown boundary data, it is necessary to know the behavior of the fundamental solutions when $r = |\mathbf{y} - \mathbf{x}|$ tends to zero, i.e., when an integration point **x** approaches a collocation point **y**. Six of the eight fundamental solutions, four in **G** and four calculated by equations (6.40), are singular. The order of their singularity can be determined by series representations. This leads to

$$\hat{P}^s_i = \mathcal{O}\left(r^0\right) \tag{6.52a}$$

$$\hat{U}^f_i = \mathcal{O}\left(r^0\right) \tag{6.52b}$$

$$\hat{U}^s_{ij} = \underbrace{\frac{1+\nu}{8\pi E(1-\nu)}\left\{r_{,i}r_{,j} + \delta_{ij}(3-4\nu)\right\}\frac{1}{r}}_{\text{elastostatic fundamental solution}} + \mathcal{O}\left(r^0\right) \tag{6.52c}$$

$$\hat{P}^f = \frac{\rho_f s}{4\pi\beta}\frac{1}{r} + \mathcal{O}\left(r^0\right) \tag{6.52d}$$

$$\hat{Q}^s_j = \frac{1+\nu}{8\pi E(1-\nu)}\left\{\alpha(1-2\nu)(r_{,n}r_{,j} - n_j) - 2\beta(1-\nu)(r_{,n}r_{,j} + n_j)\right\}\frac{1}{r} + \mathcal{O}\left(r^0\right) \tag{6.52e}$$

$$\hat{T}^f_i = \frac{\rho_f s^2}{8\pi\beta}\left\{(\alpha-\beta)\frac{1-2\nu}{1-\nu}r_{,i}r_{,n} + n_i\frac{\alpha+\beta(1-2\nu)}{1-\nu}\right\}\frac{1}{r} + \mathcal{O}\left(r^0\right) \tag{6.52f}$$

$$\hat{T}^s_{ij} = \underbrace{\frac{-\left[(1-2\nu)\delta_{ij} + 3r_{,i}r_{,j}\right]r_{,n} + (1-2\nu)(r_{,j}n_i - r_{,i}n_j)}{8\pi(1-\nu)\;r^2}}_{\text{elastostatic fundamental solution}} + \mathcal{O}\left(r^0\right) \tag{6.52g}$$

$$\hat{Q}^f = \underbrace{-\frac{r_{,n}}{4\pi r^2}}_{\text{acoustic fundamental solution}} + \mathcal{O}\left(r^0\right)\,. \tag{6.52h}$$

In equations (6.52), it is shown that the fundamental solutions are either regular (6.52a) and (6.52b), weakly singular (6.52c) - (6.52f), or strongly singular (6.52g) and (6.52h). The strong singular parts in the kernel functions (6.52g) and (6.52h) are known from elastostatics and acoustics, respectively. Therefore, shifting in (6.39) load point **y** to the boundary Γ results in the boundary integral equation

$$\int_\Gamma \begin{bmatrix} \hat{U}^s_{ij} & -\hat{P}^s_j \\ \hat{U}^f_i & -\hat{P}^f \end{bmatrix} \begin{bmatrix} \hat{t}_i \\ \hat{q} \end{bmatrix} \mathrm{d}\Gamma = \oint_\Gamma \begin{bmatrix} \hat{T}^s_{ij} & \hat{Q}^s_j \\ \hat{T}^f_i & \hat{Q}^f \end{bmatrix} \begin{bmatrix} \hat{u}_i \\ \hat{p} \end{bmatrix} \mathrm{d}\Gamma + \begin{bmatrix} c_{ij} & 0 \\ 0 & c \end{bmatrix} \begin{bmatrix} \hat{u}_i \\ \hat{p} \end{bmatrix} \tag{6.53}$$

with the integral free terms c_{ij} and c known from elastostatics and acoustics, respectively. A transformation to time domain gives, finally, the time-dependent integral equation for poroelasticity

$$\int_0^t \int_\Gamma \begin{bmatrix} U_{ij}^s(t-\tau,\mathbf{y},\mathbf{x}) & -P_j^s(t-\tau,\mathbf{y},\mathbf{x}) \\ U_i^f(t-\tau,\mathbf{y},\mathbf{x}) & -P^f(t-\tau,\mathbf{y},\mathbf{x}) \end{bmatrix} \begin{bmatrix} t_i(\tau,\mathbf{x}) \\ q(\tau,\mathbf{x}) \end{bmatrix} \mathrm{d}\Gamma \mathrm{d}\tau =$$

$$\int_0^t \oint_\Gamma \begin{bmatrix} T_{ij}^s(t-\tau,\mathbf{y},\mathbf{x}) & Q_j^s(t-\tau,\mathbf{y},\mathbf{x}) \\ T_i^f(t-\tau,\mathbf{y},\mathbf{x}) & Q^f(t-\tau,\mathbf{y},\mathbf{x}) \end{bmatrix} \begin{bmatrix} u_i(\tau,\mathbf{x}) \\ p(\tau,\mathbf{x}) \end{bmatrix} \mathrm{d}\Gamma \mathrm{d}\tau + \begin{bmatrix} c_{ij}(\mathbf{y}) & 0 \\ 0 & c(\mathbf{y}) \end{bmatrix} \begin{bmatrix} u_i(t,\mathbf{y}) \\ p(t,\mathbf{y}) \end{bmatrix} . \tag{6.54}$$

6.3.2 Boundary element formulation

A boundary element formulation is achieved following the same steps as in elasto- or viscoelastodynamics. First, the boundary surface Γ is discretized by E iso-parametric elements Γ_e where F polynomial shape functions $N_e^f(\mathbf{x})$ are defined. Hence, the following ansatz functions are used with the time-dependent nodal values $u_i^{ef}(t), t_i^{ef}(t), p^{ef}(t)$ and $q^{ef}(t)$

$$\begin{aligned} u_i(\mathbf{x},t) &= \sum_{e=1}^{E}\sum_{f=1}^{F} N_e^f(\mathbf{x})\, u_i^{ef}(t) \quad & t_i(\mathbf{x},t) &= \sum_{e=1}^{E}\sum_{f=1}^{F} N_e^f(\mathbf{x})\, t_i^{ef}(t) \\ p(\mathbf{x},t) &= \sum_{e=1}^{E}\sum_{f=1}^{F} N_e^f(\mathbf{x})\, p^{ef}(t) \quad & q(\mathbf{x},t) &= \sum_{e=1}^{E}\sum_{f=1}^{F} N_e^f(\mathbf{x})\, q^{ef}(t) \, . \end{aligned} \tag{6.55}$$

In equations (6.55), the shape functions of all four variables are denoted by the same function $N_e^f(\mathbf{x})$ indicating the same approximation level of all variables. This is not mandatory but usual. Inserting these ansatz functions (6.55) in the time dependent integral equation (6.54) yields

$$\begin{aligned} \begin{bmatrix} c_{ij}(\mathbf{y}) & 0 \\ 0 & c(\mathbf{y}) \end{bmatrix} \begin{bmatrix} u_i(\mathbf{y},t) \\ p(\mathbf{y},t) \end{bmatrix} = \sum_{e=1}^{E}\sum_{f=1}^{F} \Bigg\{ & \int_\Gamma \begin{bmatrix} U_{ij}^s(r,t) & -P_j^s(r,t) \\ U_i^f(r,t) & -P^f(r,t) \end{bmatrix} N_e^f(\mathbf{x})\, \mathrm{d}\Gamma * \begin{bmatrix} t_i^{ef}(t) \\ q^{ef}(t) \end{bmatrix} \\ & - \oint_\Gamma \begin{bmatrix} T_{ij}^s(r,t) & Q_j^s(r,t) \\ T_i^f(r,t) & Q^f(r,t) \end{bmatrix} N_e^f(\mathbf{x})\, \mathrm{d}\Gamma * \begin{bmatrix} u_i^{ef}(t) \\ p^{ef}(t) \end{bmatrix} \Bigg\} . \end{aligned} \tag{6.56}$$

Next, a time discretization has to be introduced. Since no time-dependent fundamental solutions are known, the convolution quadrature method is the most effective method compared to the possibility inverting the Laplace domain fundamental solutions at every collocation point in every time step using a series expansion [47].

Hence, after dividing time period t in N intervals of equal duration Δt, so that $t = N\Delta t$, the convolution integrals between the fundamental solutions and the nodal values in (6.56) are approximated by the convolution quadrature method, i.e., the quadrature formula (2.18) is applied to the integral equation (6.56). This results in the following boundary element time stepping formulation for $n = 0, 1, \dots, N$

$$\begin{bmatrix} c_{ij}(\mathbf{y}) & 0 \\ 0 & c(\mathbf{y}) \end{bmatrix} \begin{bmatrix} u_i(\mathbf{y}, n\Delta t) \\ p(\mathbf{y}, n\Delta t) \end{bmatrix} = \sum_{e=1}^{E} \sum_{f=1}^{F} \sum_{k=0}^{n} \left\{ \begin{bmatrix} \omega_{n-k}^{ef}\left(\hat{U}_{ij}^{s}, \mathbf{y}, \Delta t\right) & -\omega_{n-k}^{ef}\left(\hat{P}_{j}^{s}, \mathbf{y}, \Delta t\right) \\ \omega_{n-k}^{ef}\left(\hat{U}_{i}^{f}, \mathbf{y}, \Delta t\right) & -\omega_{n-k}^{ef}\left(\hat{P}^{f}, \mathbf{y}, \Delta t\right) \end{bmatrix} \begin{bmatrix} t_i^{ef}(k\Delta t) \\ q^{ef}(k\Delta t) \end{bmatrix} - \begin{bmatrix} \omega_{n-k}^{ef}\left(\hat{T}_{ij}^{s}, \mathbf{y}, \Delta t\right) & \omega_{n-k}^{ef}\left(\hat{Q}_{j}^{s}, \mathbf{y}, \Delta t\right) \\ \omega_{n-k}^{ef}\left(\hat{T}_{i}^{f}, \mathbf{y}, \Delta t\right) & \omega_{n-k}^{ef}\left(\hat{Q}^{f}, \mathbf{y}, \Delta t\right) \end{bmatrix} \begin{bmatrix} u_i^{ef}(k\Delta t) \\ p^{ef}(k\Delta t) \end{bmatrix} \right\} \tag{6.57}$$

with the integration weights corresponding to (2.22), e.g.,

$$\omega_n^{ef}\left(\hat{U}_{ij}^{s}, \mathbf{y}, \Delta t\right) = \frac{\mathscr{R}^{-n}}{L} \sum_{\ell=0}^{L-1} \int_{\Gamma} \hat{U}_{ij}^{s}\left(\mathbf{x}, \mathbf{y}, \frac{\gamma\left(\mathscr{R} e^{i\ell\frac{2\pi}{L}}\right)}{\Delta t}\right) N_e^f(\mathbf{x})\, d\Gamma\, e^{-in\ell\frac{2\pi}{L}} . \tag{6.58}$$

Note, the calculation of the integration weights is only based on the Laplace transformed fundamental solutions which are available. Therefore, with the time stepping procedure (6.57) a boundary element formulation for poroelastodynamics is given without time-dependent fundamental solutions.

To calculate the integration weights ω_{n-k}^{ef} in (6.57), spatial integration over the boundary Γ has to be performed. Because the essential constituents of the Laplace transformed fundamental solutions are exponential functions, i.e., the integrand is smooth, the regular integrals are evaluated by standard Gaussian quadrature rule. The weakly singular parts of the integrals in (6.57) are regularized by polar coordinate transformation. The strongly singular integrals in (6.57) are equal to those of elastostatics or acoustics, respectively, and, hence, the regularization methods known from these theories can be applied, e.g., the method suggested by Guiggiani and Gigante [101]. Moreover, to obtain for equation (6.57) a system of algebraic equations, collocation is used at every node of the shape functions $N_e^f(\mathbf{x})$. Now, identical as in visco- and elastodynamics, a recursion formula ($m = n - k$)

$$\omega_0(\mathbf{C})\,\mathbf{d}^n = \omega_0(\mathbf{D})\,\bar{\mathbf{d}}^n + \sum_{m=1}^{n} \left(\omega_m(\mathbf{U})\,\mathbf{t}^{n-m} - \omega_m(\mathbf{T})\,\mathbf{u}^{n-m}\right) \qquad n = 1, 2, \dots, N \tag{6.59}$$

using the same matrices and vectors as in equation (4.22) is established and, finally, a direct equation solver is applied.

6.4 Numerical studies

In order to validate the proposed poroelastic boundary element approach, two problems are investigated: First, the influence of the time step size and the mesh size is analyzed by comparing the approximated results achieved by the BEM to an analytical solution of a 1-d column, and, second, a half space under a vertical load is considered for studying wave propagation in different material modelings.

In the following tests, the underlying multistep method $\gamma(z)$ is a BDF 2 and $L = N$ is chosen due to the experiences in the visco- and elastodynamic formulations. Further, numerical stable solutions are only achieved if dimensionless variables are introduced. As suggested in [47], the dimensionless spatial and temporal variables are

$$x_i \to \frac{x_i}{\rho\kappa V} \qquad t \to \frac{t}{\rho\kappa} \qquad \text{with} \qquad V = \sqrt{\frac{K + \frac{4}{3}G + \alpha^2 \frac{R}{\phi^2}}{\rho}} \tag{6.60}$$

where the velocity V is the compression wave velocity of a poroelastic solid with an inviscid interstitial fluid. These non-dimensional variables are connected with dimensionless material parameters

$$\begin{aligned} K &\to \frac{K}{K + \frac{4}{3}G + \alpha^2 \frac{R}{\phi^2}} \qquad & G &\to \frac{G}{K + \frac{4}{3}G + \alpha^2 \frac{R}{\phi^2}} \qquad & R &\to \frac{R}{K + \frac{4}{3}G + \alpha^2 \frac{R}{\phi^2}} \\ \kappa &\to \kappa = 1 & \rho &\to \frac{\rho}{\rho} = 1 & \rho_f &\to \frac{\rho_f}{\rho} \end{aligned} \tag{6.61}$$

The condition number of the equation system confirms the necessity of dimensionless variables. It has an order of 10^9 using the dimensionless variables instead of 10^{22} else.

6.4.1 Influence of time step size and mesh size

The one dimensional (1-d) column of length 3 m as sketched in Fig. 6.1 is considered. It is assumed that the side walls and the bottom are rigid, frictionless, and impermeable. Hence, the displacements normal to the surface are blocked and the column is free to slide only parallel to the wall. At the top, the total stress vector $t_y = -1\,\mathrm{N/m^2}\,H(t)$ and the pore pressure $p = 0\,\mathrm{N/m^2}$ is a given, i.e., a normal pressure force of intensity one starts acting with $t > 0\,\mathrm{s}$ and fluid particles are assumed to be on a free fluid surface. Due to these restrictions, the 3-d continuum is reduced to a 1-d continuum with the only degree of freedom in y-direction. This 1-d problem has been analytically solved (see Sect. 7.1) and its result is compared to the boundary element solution for a 3-d rod ($3\,\mathrm{m} \times 1\,\mathrm{m} \times 1\,\mathrm{m}$). The boundary conditions are modeled as above.

Results achieved with three different meshes are compared to study convergence and sensitivity on spatial discretization. The discretizations are those used in the corresponding study of visco- and elastodynamic formulations (see Fig. 4.4): a very

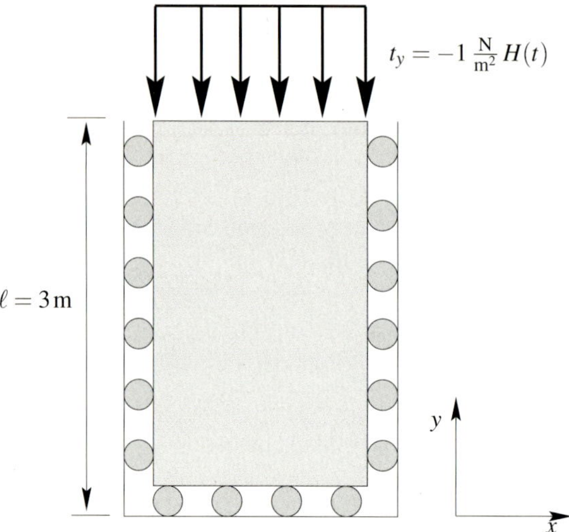

Fig. 6.1. Geometry and dynamic loading of one-dimensional column

coarse mesh with 56 linear triangles on 30 nodes (mesh 1), a finer mesh with 112 linear triangles on 58 nodes (mesh 2), and a non uniform mesh with refinement towards the edges with 324 linear triangles on 164 nodes (mesh 3). The material properties are those corresponding to Berea sandstone (see Table 7.1) except that zero Poisson's ratio is set to model the 1-d behavior. This results in $G = 7.2 \cdot 10^9\,\mathrm{N/m^2}$ and $K = 4.8 \cdot 10^9\,\mathrm{N/m^2}$, respectively.

In previous chapters, the dimensionless value β was introduced to study the dependency on time step size for different spatial discretizations. The elastodynamic compression wave velocity c_1 is used in the definition of β (4.25). Contrary, in poroelasticity the wave velocities are time-dependent, and, therefore, only an approximate value can be used similar to viscoelastodynamics where the wave velocity defined with the initial moduli is taken. Here, the compression wave velocity V (6.60) corresponding to an inviscid interstitial fluid is chosen leading to

$$\beta = \frac{V\Delta t}{r_e} \tag{6.62}$$

with r_e defined in Sect. 4.3.

In Fig. 6.2, the vertical displacement on top of the column and, in Fig. 6.3, the traction at the bottom is depicted versus time for three meshes. An optimal time step size is used for every mesh, respectively. As expected, the results achieved with mesh 3 are the closest to the analytical solution, whereas the coarse mesh 1 shows a phase shift towards larger times and numerical damping. The results for mesh 2 are between those of mesh 1 and mesh 3. The traction solution indicates that mesh 1 is not suitable for this calculation, whereas mesh 2 seems to be acceptable. But the pressure solution, not presented in Figs. 6.2 or 6.3, indicates that even mesh 2 is not sufficient for an accurate calculation of the pore pressure.

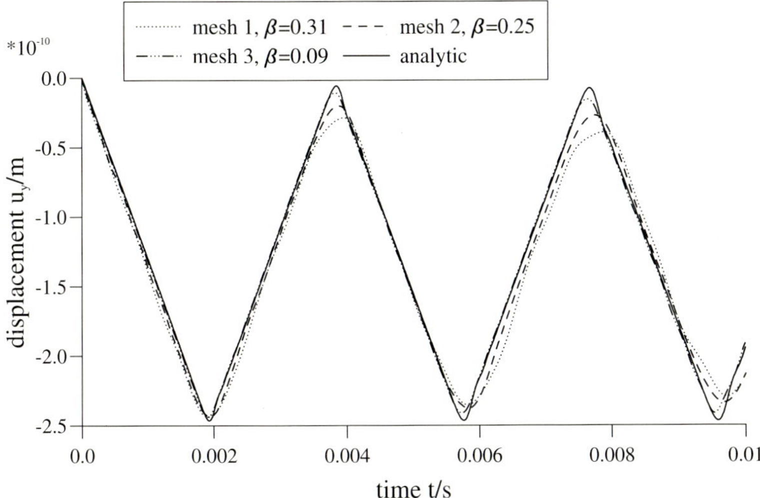

Fig. 6.2. Longitudinal displacement at the top of the column versus time: Influence of mesh size

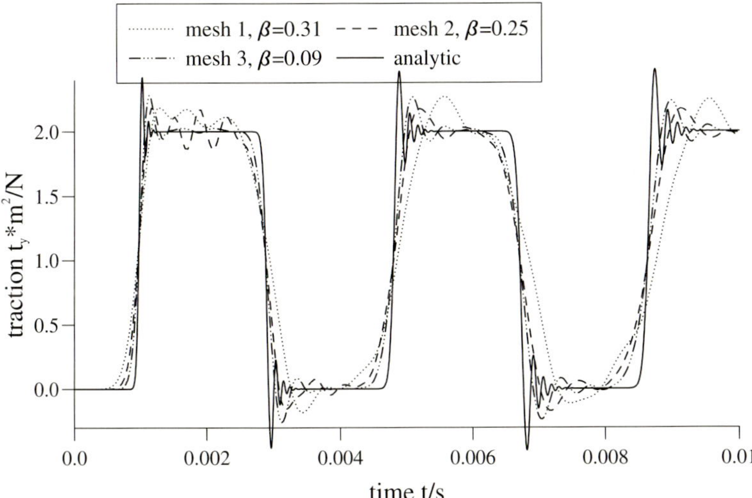

Fig. 6.3. Traction at the bottom of the column versus time: Influence of mesh size

Next, the influence of the time step size is studied using mesh 3 and the material data of Berea sandstone. In Fig. 6.4, the vertical displacement at top of the column is plotted versus time for different time step sizes of the numerical BEM model. Additionally, in Fig. 6.5, the pore pressure at the bottom of the column is shown. Clearly, as in all boundary element time stepping procedures, a critical time step size is observed below which the results become unstable. In the presented solu-

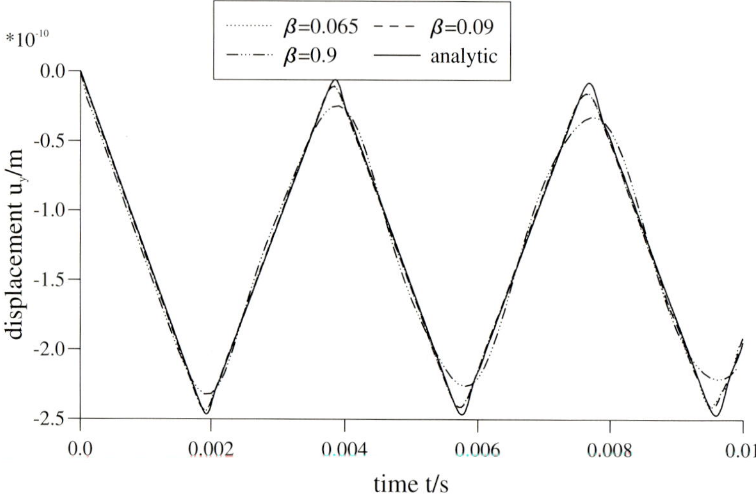

Fig. 6.4. Longitudinal displacement at the top of the column versus time: Influence of time step size

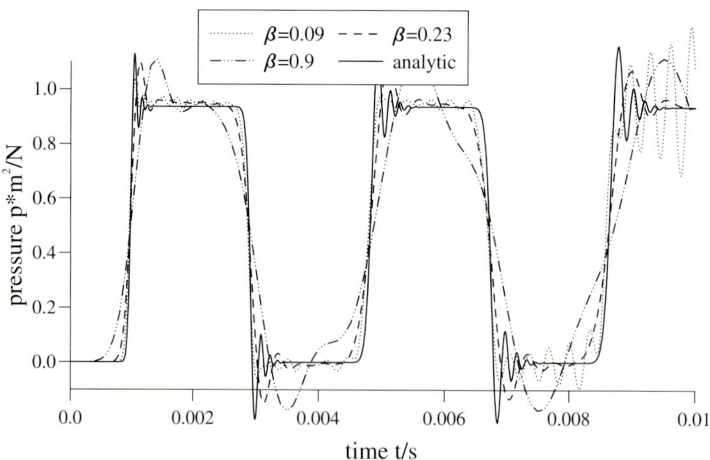

Fig. 6.5. Pore pressure at the bottom of the column versus time: Influence of time step size

tion, these instabilities occur for very small time step sizes $\beta = 0.065$ only in the more sensitive pressure solution. Indeed, the displacement solution does not show any instabilities within the presented time window. Stable results for $\beta > 0.09$, i.e., $\Delta t > 0.00001$ s, are sufficient close to the analytical solution and show only slight dependencies on the time step size. The latter statement is concluded since the results for $\Delta t = 0.00001$ s and $\Delta t = 0.00002$ s (not presented in Figs. 6.4 or 6.5) are almost identical in spite of the doubled time step size. Due to their good agreement they would not be distinguishable in a plot. In the long time behavior, a discrepancy in the maximum values of the analytical 1-d solution and the numerical 3-d solution is observed. Concluding this study, it can be stated that the proposed method works well, even for small time step sizes if sufficient fine spatial discretization is used. Then the method is nearly independent on the time step size if $\beta < 1$ is regarded.

A remark on sufficient spatial discretizations should be added. Comparing the study concerning spatial and temporal discretization for the elastodynamic (Figs. 4.5 – 4.8) or viscoelastodynamic formulation (Figs. 5.4 – 5.7) with the behavior presented here, it can be concluded that the poroelastodynamic formulation needs finer meshes to achieve qualitatively the same results as a comparable elasto- or viscoelastodynamic calculation. This is caused by modeling the interaction between the solid skeleton and the interstitial fluid in the used poroelastic theory. Due to this, two different time scales one for the solid and one for the fluid have to be reproduced by one method, i.e., the spatial and temporal discretization can not be optimal for both parts of the system.

The eigenfrequencies of the poroelastic column visualize this fact more clearly. In Fig. 6.6, the absolute value of the displacement at the top of the column is plotted versus frequency for Berea sandstone using the analytical solution. To show the influence of the interstitial fluid, the frequency response of the column modeled

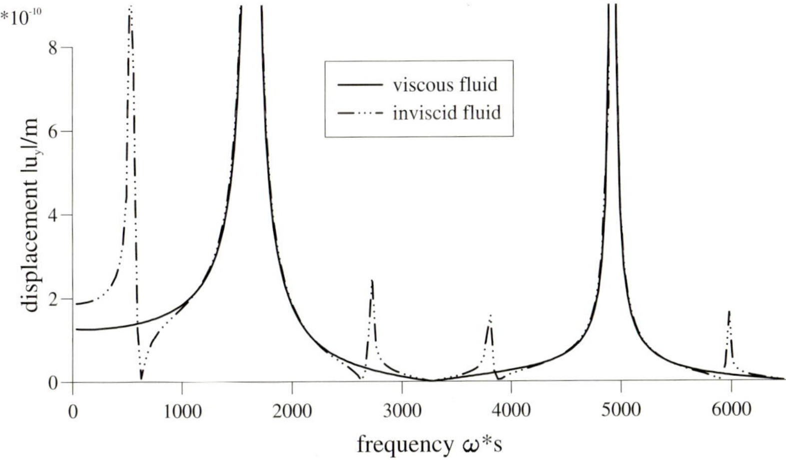

Fig. 6.6. Longitudinal displacement at the top of the column versus frequency: Comparison viscous and inviscid fluid

with a viscous fluid ($\kappa = 1.9 \cdot 10^{-10}\,\mathrm{m}^4/(\mathrm{Ns})$ as given in Table 7.1) is compared to an inviscid fluid ($\kappa = 1.9\,\mathrm{m}^4/(\mathrm{Ns})$), i.e., the friction between the solid and the interstitial fluid is neglected. This friction introduces damping in the system, which does not change the resonance frequencies, but reduce their amplitudes. This effect is observed in Fig. 6.6. There, in the realistic damped case ($\kappa = 1.9 \cdot 10^{-10}\,\mathrm{m}^4/(\mathrm{Ns})$) the first two eigenfrequencies of the solid skeleton are visible. In the undamped case more resonance peaks are visible. These must be the resonance frequencies of the interstitial fluid which are suppressed by damping in the realistic case. Now, in a poroelastic calculation spatial and temporal discretization must be sufficient for both constituents, which lead in general to finer meshes than needed to model single phase materials.

6.4.2 Poroelastic half space

Next, a poroelastic half space using material data of soil (see Table 7.1) is considered. The discretization is truncated behind 4 m around the center where 684 linear triangles on 397 nodes are used (see Fig. 6.7). The half space is loaded by a total stress $t_z = -1\,\mathrm{N/m^2}\,H(t)$ and is taken to be permeable (i.e., $p = 0\,\mathrm{N/m^2}$) in area A (shaded in Fig. 6.7, 2 m radius), whereas the remaining surface (not shaded) is as-

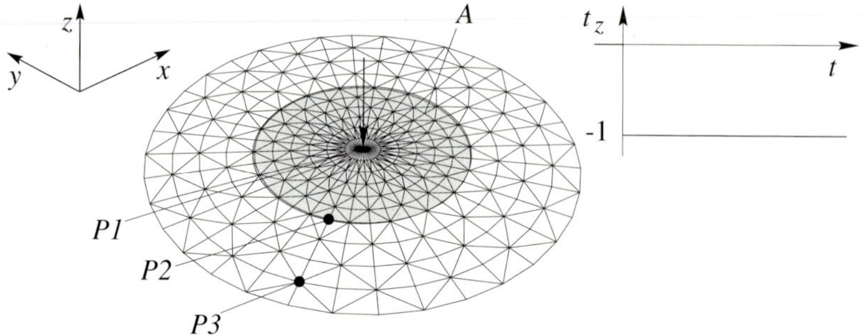

Fig. 6.7. Half space under vertical load: Discretization, loading area A, and load history

sumed to be impermeable and traction free, i.e., zero flux and zero traction is given.

First, the time history of the pore pressure and of the displacement at point $P3$ (3.5 m distance from the center) is presented. In Fig. 6.8, both are depicted with different scales for each. The time, when the first compressional wave arrives, is clearly identified ($t \approx 0.0008\,\mathrm{s}$). Then, the surface lifts and causes a negative pressure in the interior because water is sucked in this area from the surrounding. After the arrival of shear wave and Rayleigh wave the surface starts to move down and consequently the pressure is increased. The arrival time of the shear wave ($t \approx 0.0056\,\mathrm{s}$) and of the Rayleigh wave ($t \approx 0.006\,\mathrm{s}$) can not be distinguished on the plot because they are too close to each other in this short distance from the excitation. Finally, when the displacement has reached its maximum value the pressure solution starts to decrease.

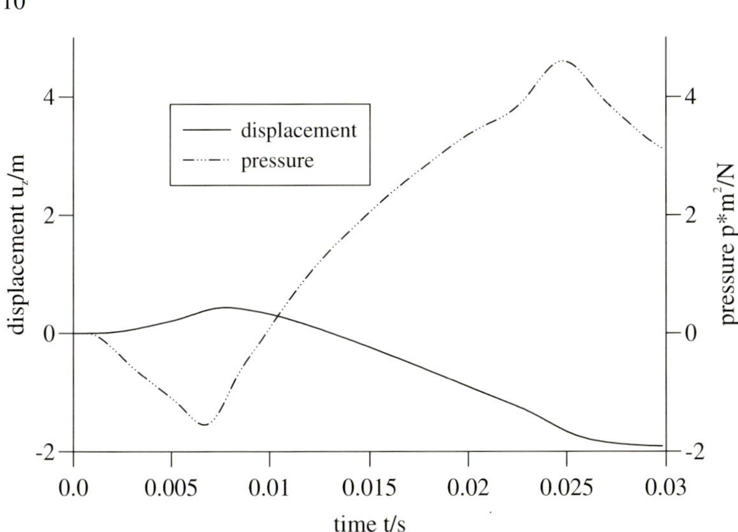

Fig. 6.8. Pore pressure and displacement versus time at point $P3$

To validate the proposed poroelastic boundary element formulation and to get an impression of the influence of the interstitial fluid in a poroelastic modeled soil, the time history of the displacement at the points $P1$, $P2$, and $P3$ (see mesh in Fig. 6.7) is compared for the poroelastic soil and for two other elastic modeled soils with the same shear modulus as the poroelastic medium, but one with its undrained Poisson's ratio ($\nu_u = 0.49$) and the other with its drained Poisson's ratio ($\nu = 0.298$, see Table 7.3). These two elastic modeled media represent a possibility to model a water saturated soil as a one-phase material where the undrained case assumes a strong influence of the interstitial fluid and the drained case less influence. Therefore, results for the two elastic media should give an upper and a lower bound for the poroelastic soil.

The influence of the interstitial fluid becomes obvious by changing the boundary condition in the not shaded area (Fig. 6.7) from impermeable, i.e., zero flux boundary boundary condition, to permeable, i.e., zero pressure boundary condition. For both, in Fig. 6.9 and Fig. 6.10, the vertical displacement versus time is depicted for all three material models and all points. In Fig. 6.9, a permeable surface is assumed, whereas in Fig. 6.10, an impermeable surface is assumed.

First, effects of the permeable boundary condition (Fig. 6.9) are discussed. At the center (point $P1$), exactly at the point of excitation, the absolute displacement value starts to increase beginning at time $t = 0\,\mathrm{s}$ until the final value is reached ($t \approx 0.01\,\mathrm{s}$). For the poroelastic material, this increase is slower than those for the elastic materials. However, approaching the final state the displacement of the poroelastic material is in between those of the elastic media, which confirms that the drained and undrained elastic material represent the extrema of interstitial fluid effects, i.e.,

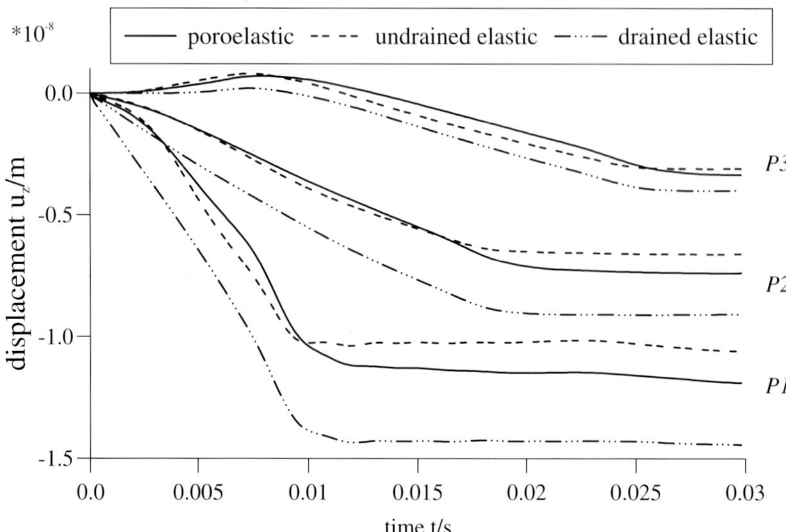

Fig. 6.9. Vertical displacement versus time at three locations $P1, P2$, and $P3$: Permeable boundary condition at the soil surface

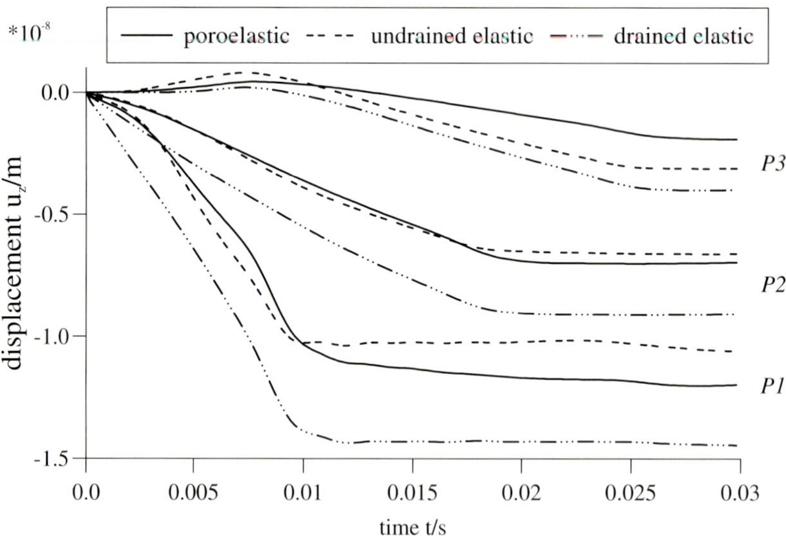

Fig. 6.10. Vertical displacement versus time at three locations $P1, P2$, and $P3$: Impermeable boundary condition at the soil surface

strong influence and nearly no influence, as stated above. This is observed as well at the points $P2$ and $P3$. At point $P2$, the poroelastic solution tends to the undrained elastic solution. This tendency is even stronger at point $P3$ outside the loading area.

A physical interpretation is found by the interstitial fluid. The flow of this fluid is modeled viscous and, thus, the friction between the interstitial fluid and the solid skeleton causes a resistance of the material. However, in the permeable case, the water can flow out of the surface and, therefore, causes only a small resistance. In contrast, an impermeable surface prevents the water from escaping through the surface, instead the fluid is pressed in the surrounding area. Due to the viscous flow, this results in a strong resistance, i.e., the material responds more stiff. This effect is clearly observed in Fig. 6.10 at point $P3$. At this point, both elastic materials are no longer any bound for the poroelastic solution. Instead, the absolute displacement values for the poroelastic medium are smaller as the values for both elastic materials. Physically, the poroelastic material behaves more stiff than the elastic material models. At point $P1$, the boundary condition does not show any impact on the displacement.

To confirm the statements above, in Fig. 6.11 the displacement at the location $P1$ and in Fig. 6.12 the displacement at the location $P3$ is compared for an inviscid fluid and an viscous fluid with permeable and impermeable boundary condition. As above, the influence of the boundary condition is clearly demonstrated. At the center (point $P1$), just below the excitation, the boundary condition has no influence contrary to point $P3$. There, the results differ in amplitude. The impermeable boundary condition leads to much smaller displacement values than the permeable boundary condition, i.e., a stiffer half space is modeled.

Further, the effect of the friction between the interstitial fluid and the solid skeleton is visualized. At point $P1$, the poroelastic soil with the inviscid fluid behaves like the drained elastic medium, i.e., the fluid has no influence on the response. Contrary, the poroelastic soil with a viscous fluid behaves in short time ($t < 0.01\,\mathrm{s}$) like the undrained elastic soil, i.e., the fluid has a strong influence. Far away, at point $P3$, these differences are no longer so strong because the material had time to "relax". There, the boundary condition has more influence than the viscosity of the fluid.

To the author best knowledge, for the given poroelastic half space problem no analytical solution exists. Therefore, as in the viscoelastic case, a final check will use a boundary element calculation in Laplace domain with subsequent inverse transformation for comparison. Again, the method of Durbin is used with the parameter α determined by $\alpha T = 5$ as given in the original work [79]. There, $T = 0.8 \cdot t_{max}$ was defined leading here to: $T = 0.024\,\mathrm{s}$ and $\alpha = 200$. In Fig. 6.13, the pressure at point $P3$ calculated with the proposed time-dependent formulation (CQM) is compared to the solution achieved via inverse transformation (Durbin). Except for some differences at time $t \approx 0.006\,\mathrm{s}$ and at time $t \approx 0.025\,\mathrm{s}$, both solutions agree quite well. The corresponding displacement solutions would cover each other so that the worst case is presented. This shows the reliability of the proposed formulation. However, contrary to the viscoelastic case, for this poroelastic problem the Durbin method shows as good results as the proposed formulation. But, a simple change of the loading

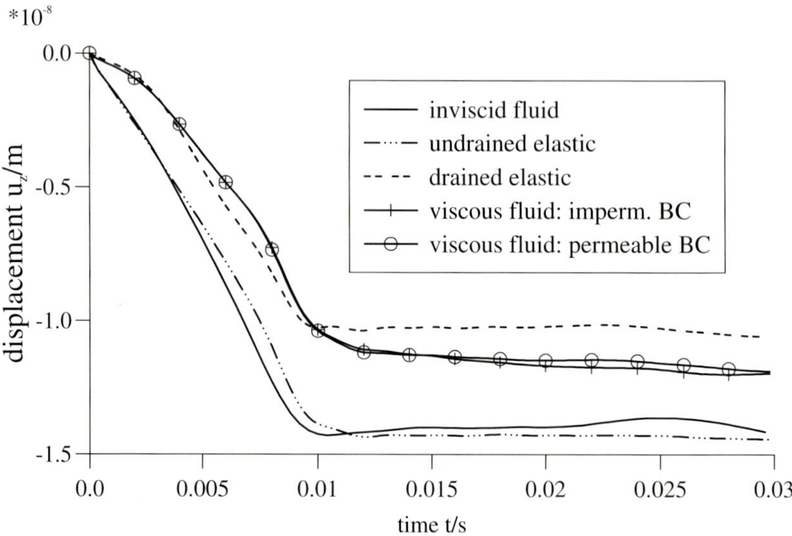

Fig. 6.11. Vertical displacement at point $P1$ versus time: Influence of boundary condition and viscosity of the interstitial fluid

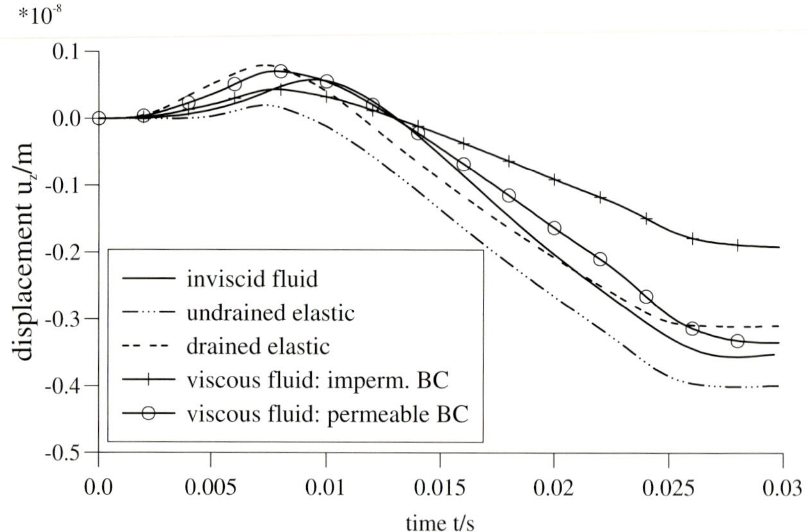

Fig. 6.12. Vertical displacement at point $P3$ versus time: Influence of boundary condition and viscosity of the interstitial fluid

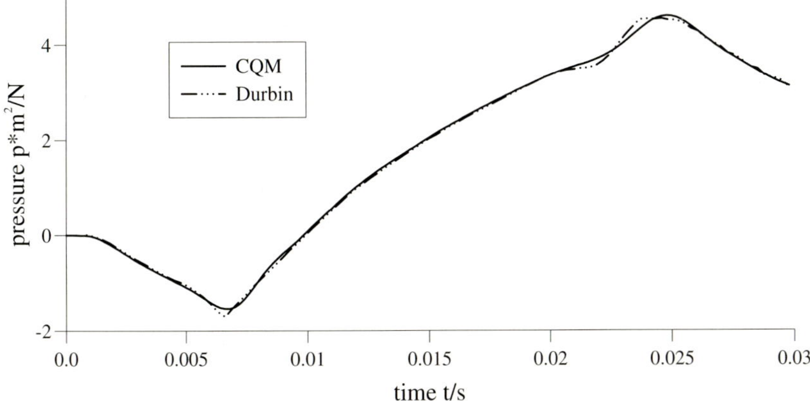

Fig. 6.13. Pressure at point $P3$ versus time: Comparison of the proposed formulation (CQM) with a formulation performed in Laplace domain with subsequent inverse transformation (Durbin)

time history to a "hat" function causes in the Laplace domain solution problems in finding the right parameter α as shown in the viscoelastic case. Also, no contact problem or other non-linear boundary condition can be treated by the Laplace domain formulation. So, totally, the proposed formulation has to be preferred.

Concluding this study, both the reliability of the proposed formulation and the importance of modeling saturated soil as a poroelastic two-phase continuum has been demonstrated. Further, the influence of correctly modeling the boundary condition of the surface has been shown. These results are valid for other poroelastic materials, too.

7. Wave propagation

The study of wave propagation phenomena in different materials give necessary insight to construct buildings or prevent houses for destructive or only disturbing tremors. These, in general, 3-d problems can only be treated by numerical methods as shown in the previous chapters. For some simplified 1-d problems analytical or semi-analytical solutions exist. These solutions are used to either control numerical methods or also to study some basic effects of wave propagation. In this chapter, first, wave propagation in a poroelastic 1-d column is examined, followed by 3-d calculations using the proposed boundary element formulations.

7.1 Wave propagation in poroelastic one-dimensional column

7.1.1 Analytical solution

A one-dimensional column of length ℓ as sketched in Fig. 7.1 is considered. It is assumed that the side walls and the bottom are rigid, frictionless, and impermeable. Hence, the displacements normal to the surface are blocked and the column is otherwise free to slide parallel to the wall. At the top, the stress σ_y and the pressure p

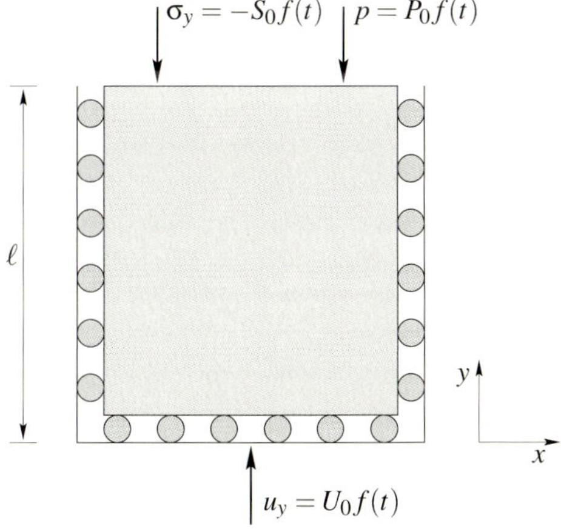

Fig. 7.1. One-dimensional column under dynamic loading

are prescribed. Due to these restrictions only the displacement u_y and the pore pressure p remain as degrees of freedom. This one-dimensional example can be used to study the influence of poroelastic parameters on wave propagation, or it can be seen as an approximation of a poroelastic half-space by setting the layer depth ℓ large. Here, the study will be focused on observing the interplay of the two compressional waves.

For this, the governing set of differential equations (6.10) and (6.11) is reduced to two scalar coupled ordinary differential equations

$$E\hat{u}_{y,yy} - (\alpha-\beta)\,\hat{p}_{,y} - s^2\,(\rho-\beta\rho_f)\,\hat{u}_y = 0 \tag{7.1}$$

$$\frac{\beta}{s\rho_f}\hat{p}_{,yy} - \frac{\phi^2 s}{R}\hat{p} - (\alpha-\beta)\,s\hat{u}_{y,y} = 0\,, \tag{7.2}$$

with the modulus $E = K + 4/3G$ and vanishing body forces F_i and sources a. The boundary conditions in Laplace domain are

$$\begin{aligned} &\hat{u}_y\,(y=0) = U_0, \qquad &&\hat{q}_y\,(y=0) = 0 \quad \text{and} \\ &\hat{\sigma}_y\,(y=\ell) = -S_0, \qquad &&\hat{p}\,(y=\ell) = P_0\,, \end{aligned} \tag{7.3}$$

where an impulse function for the temporal behavior $f(t) = \delta(t)$ is assumed, together with vanishing initial conditions. Each of the non-zero boundary conditions in (7.3) represents a different type of loading. Due to the neglected body forces this is a system of homogeneous ordinary differential equations with inhomogeneous boundary conditions. Such a system can be solved by the following exponential ansatz

$$\hat{u}_y\,(y) = U\mathrm{e}^{\lambda s y} \qquad \hat{p}\,(y) = P\mathrm{e}^{\lambda s y}\,. \tag{7.4}$$

Inserting the ansatz functions (7.4) in equations (7.1) and (7.2) results in an Eigenvalue problem for λ

$$\begin{bmatrix} E\lambda^2 - (\rho-\beta\rho_f) & -(\alpha-\beta)\dfrac{\lambda}{s} \\ -s\,(\alpha-\beta)\,\lambda & \lambda^2\dfrac{\beta}{\rho_f} - \dfrac{\phi^2}{R} \end{bmatrix} \begin{bmatrix} U \\ P \end{bmatrix} = \mathbf{0}\,, \tag{7.5}$$

with the characteristic equation

$$\underbrace{E\frac{\beta}{\rho_f}}_{A}\lambda^4 - \underbrace{\left(E\frac{\phi^2}{R} + (\rho-\beta\rho_f)\frac{\beta}{\rho_f} + (\alpha-\beta)^2\right)}_{B}\lambda^2 + \underbrace{\frac{\phi^2\,(\rho-\beta\rho_f)}{R}}_{C} \stackrel{!}{=} 0\,. \tag{7.6}$$

The characteristic equation (7.6) has the four complex roots

$$\lambda_1 = -\lambda_3 = \sqrt{\frac{B+\sqrt{B^2-4AC}}{2A}} \qquad \lambda_2 = -\lambda_4 = \sqrt{\frac{B-\sqrt{B^2-4AC}}{2A}}\,. \tag{7.7}$$

This leads to the complete solution of the homogeneous problem

$$\hat{u}_y(y) = \sum_{i=1}^{4} U_i \mathrm{e}^{\lambda_i s y} \qquad \hat{p}(y) = \sum_{i=1}^{4} P_i \mathrm{e}^{\lambda_i s y} . \tag{7.8}$$

The eight unknown constants U_i and P_i, $i = 1, \ldots, 4$, can not be determined by the four boundary conditions (7.3) alone. Also none of the complex roots can be excluded due to physical reasons. But the Eigenvector of the system (7.5) gives the relation

$$P_i = \underbrace{\frac{E\lambda_i^2 - (\rho - \beta\rho_f)}{(\alpha - \beta)\lambda_i}}_{d_i} \cdot sU_i . \tag{7.9}$$

Finally, if the solution (7.8) with the property (7.9) is inserted into the one-dimensional form of the constitutive equation (6.3)

$$\hat{\sigma}_y(s,y) = E\hat{u}_{y,y} - \alpha\hat{p} = E\sum_{i=1}^{4} \lambda_i s U_i \mathrm{e}^{\lambda_i s y} - \alpha\hat{p}(s,y) \tag{7.10}$$

and the one-dimensional form of Darcy's law (6.8)

$$\begin{aligned} \hat{q}_y(s,y) &= -\frac{\beta}{s\rho_f}\left(\hat{p}_{,y} + s^2\rho_f\hat{u}_y\right) \\ &= -\frac{\beta E s}{\rho_f(\alpha - \beta)} \sum_{i=1}^{4} \lambda_i^2 U_i \mathrm{e}^{\lambda_i s y} + \beta s\left(\frac{\rho - \alpha\rho_f}{\rho_f(\alpha - \beta)}\right)\hat{u}_y(s,y) \end{aligned} \tag{7.11}$$

the remaining four constants U_i can be fit to the four boundary conditions. This leads to four equations with four unknowns

$$\begin{bmatrix} d_1\mathrm{e}^{\lambda_1\ell} & -d_1\mathrm{e}^{-\lambda_1\ell} & d_2\mathrm{e}^{\lambda_2\ell} & -d_2\mathrm{e}^{-\lambda_2\ell} \\ \lambda_1\mathrm{e}^{\lambda_1\ell} & -\lambda_1\mathrm{e}^{-\lambda_1\ell} & \lambda_2\mathrm{e}^{\lambda_2\ell} & -\lambda_2\mathrm{e}^{-\lambda_2\ell} \\ \lambda_1^2 & \lambda_1^2 & \lambda_2^2 & \lambda_2^2 \\ 1 & 1 & 1 & 1 \end{bmatrix} \begin{bmatrix} U_1 \\ U_2 \\ U_3 \\ U_4 \end{bmatrix} = \begin{bmatrix} P_0\, s^{-1} \\ (\alpha P_0 - S_0)\, (Es)^{-1} \\ U_0\,(\rho - \alpha\rho_f)\, E^{-1} \\ U_0 \end{bmatrix} \tag{7.12}$$

which can be solved, preferably, with the aid of computer algebra.

Finally, the solutions for the displacement and the pore pressure are achieved by inserting these coefficients in the ansatz functions (7.8). As the problem at hand is linear the superposition principle is valid. Therefore, the solution can be divided in different load cases. The results are for *stress boundary conditions* $\hat{u}_y(y=0)=0, \hat{\sigma}_y(y=\ell)=-S_0$ and $\hat{p}(y=\ell)=0$

$$\hat{u}_y = \frac{S_0}{E(d_1\lambda_2 - d_2\lambda_1)} \left[\frac{d_2\left(e^{-\lambda_1 s(\ell-y)} - e^{-\lambda_1 s(\ell+y)}\right)}{s\left(1+e^{-2\lambda_1 s\ell}\right)} - \frac{d_1\left(e^{-\lambda_2 s(\ell-y)} - e^{-\lambda_2 s(\ell+y)}\right)}{s\left(1+e^{-2\lambda_2 s\ell}\right)} \right] \tag{7.13}$$

$$\hat{p} = \frac{S_0 d_1 d_2}{E(d_1\lambda_2 - d_2\lambda_1)} \left[\frac{\left(e^{-\lambda_1 s(\ell-y)} + e^{-\lambda_1 s(\ell+y)}\right)}{1+e^{-2\lambda_1 s\ell}} - \frac{\left(e^{-\lambda_2 s(\ell-y)} + e^{-\lambda_2 s(\ell+y)}\right)}{1+e^{-2\lambda_2 s\ell}} \right] \tag{7.14}$$

for *pressure boundary conditions*: $\hat{u}_y(y=0)=0, \hat{\sigma}_y(y=\ell)=0$ and $\hat{p}(y=\ell)=P_0$

$$\hat{u}_y = \frac{P_0}{Es(d_1\lambda_2 - d_2\lambda_1)} \left[\frac{(E\lambda_2 - \alpha d_2)\left(e^{-\lambda_1 s(\ell-y)} - e^{-\lambda_1 s(\ell+y)}\right)}{1+e^{-2\lambda_1 s\ell}} - \frac{(E\lambda_1 - \alpha d_1)\left(e^{-\lambda_2 s(\ell-y)} - e^{-\lambda_2 s(\ell+y)}\right)}{1+e^{-2\lambda_2 s\ell}} \right] \tag{7.15}$$

$$\hat{p} = \frac{P_0}{E(d_1\lambda_2 - d_2\lambda_1)} \left[\frac{d_1(E\lambda_2 - \alpha d_2)\left(e^{-\lambda_1 s(\ell-y)} + e^{-\lambda_1 s(\ell+y)}\right)}{1+e^{-2\lambda_1 s\ell}} - \frac{d_2(E\lambda_1 - \alpha d_1)\left(e^{-\lambda_2 s(\ell-y)} + e^{-\lambda_2 s(\ell+y)}\right)}{1+e^{-2\lambda_2 s\ell}} \right] \tag{7.16}$$

and for *displacement boundary conditions*: $\hat{u}_y(y=0)=U_0, \hat{\sigma}_y(y=\ell)=0$ and $\hat{p}(y=\ell)=0$

$$\hat{u}_y = \frac{U_0}{E(\lambda_2^2 - \lambda_1^2)} \left[\frac{(E\lambda_2^2 + \alpha\rho_f - \rho)\left(e^{-\lambda_1 s(2\ell-y)} + e^{-\lambda_1 sy}\right)}{1+e^{-2\lambda_1 s\ell}} - \frac{(E\lambda_1^2 + \alpha\rho_f - \rho)\left(e^{-\lambda_2 s(2\ell-y)} + e^{-\lambda_2 sy}\right)}{1+e^{-2\lambda_2 s\ell}} \right] \tag{7.17}$$

$$\hat{p} = \frac{U_0 s}{E(\lambda_2^2 - \lambda_1^2)} \left[\frac{d_1(E\lambda_2^2 + \alpha\rho_f - \rho)\left(e^{-\lambda_1 s(2\ell-y)} - e^{-\lambda_1 sy}\right)}{1+e^{-2\lambda_1 s\ell}} - \frac{d_2(E\lambda_1^2 + \alpha\rho_f - \rho)\left(e^{-\lambda_2 s(2\ell-y)} - e^{-\lambda_2 sy}\right)}{1+e^{-2\lambda_2 s\ell}} \right] \tag{7.18}$$

The corresponding stress and flux is calculated with the constitutive equation (7.10) and Darcy's law (7.11), respectively.

Note, due to the dependence of β to the Laplace parameter s, the roots λ_i and consequently d_i are dependent of s. Therefore, an analytical inverse Laplace transform of the solutions above is in general not possible. However, if the damping due to the relative motion of the fluid and the solid is neglected, i.e., the permeability tends to infinity

$$\kappa \to \infty \qquad \Longrightarrow \qquad \beta \approx \frac{\phi^2 \rho_f}{\rho_a + \phi \rho_f} , \tag{7.19}$$

an analytical inverse Laplace transform can be found.

Special case of $\kappa \to \infty$. Under this assumption β and consequently λ_i and d_i are constant with respect to s. Then, in solutions (7.13) - (7.18) only the expressions with the exponential function are dependent on s. In the analysis of the corresponding elastic problem the same expressions appear. Following the procedure in [98], the series expansion

$$\frac{1}{1+\mathrm{e}^{-2\lambda_i s \ell}} = \sum_{n=0}^{\infty} (-1)^{-n} \mathrm{e}^{-2\ell n s \lambda_i} \tag{7.20}$$

gives

$$\begin{aligned} &\frac{\left(\mathrm{e}^{-\lambda_i s(\ell - y)} - \mathrm{e}^{-\lambda_i s(\ell + y)}\right)}{s\left(1+\mathrm{e}^{-2\lambda_i s \ell}\right)} \\ &\qquad = \sum_{n=0}^{\infty} (-1)^{-n} \left(\frac{1}{s} \mathrm{e}^{-\lambda_i s(\ell(2n+1)-y)} - \frac{1}{s} \mathrm{e}^{-\lambda_i s(\ell(2n+1)+y)} \right) . \end{aligned} \tag{7.21}$$

Now, an inverse Laplace transform is possible term by term of the series above. With the relations

$$\frac{1}{s} \mathrm{e}^{-\lambda_i s(\ell(2n+1)-y)} \bullet\!\!-\!\!\circ\, H\left(t - \lambda_i \left(\ell\,(2n+1) - y\right)\right) \tag{7.22}$$

$$\mathrm{e}^{-\lambda_i s(\ell(2n+1)-y)} \bullet\!\!-\!\!\circ\, \delta\left(t - \lambda_i \left(\ell\,(2n+1) - y\right)\right) , \tag{7.23}$$

the inverse transform of (7.13) - (7.18) is given. The response in time domain can be calculated with the convolution integral, e.g., for the displacements

$$u_y(t,y) = \int_0^t \mathscr{L}^{-1}\left\{\hat{u}_y(s,y)\right\}(\tau,y)\; f(t-\tau)\, \mathrm{d}\tau . \tag{7.24}$$

Assuming a Heaviside step function as temporal behavior of the load, i.e., $f(t) = H(t)$, the response in time domain is

$$u_y = \frac{S_0}{E(d_1\lambda_2 - d_2\lambda_1)} \sum_{n=0}^{\infty} (-1)^{-n} \Big\{ d_2 \big[(t - \lambda_1(\ell(2n+1) - y)) H(t - \lambda_1(\ell(2n+1) - y)) - (t - \lambda_1(\ell(2n+1) + y)) H(t - \lambda_1(\ell(2n+1) + y)) \big] - d_1 \big[(t - \lambda_2(\ell(2n+1) - y)) H(t - \lambda_2(\ell(2n+1) - y)) - (t - \lambda_2(\ell(2n+1) + y)) H(t - \lambda_2(\ell(2n+1) + y)) \big] \Big\} \tag{7.25}$$

$$p = \frac{S_0 d_1 d_2}{E(d_1\lambda_2 - d_2\lambda_1)} \sum_{n=0}^{\infty} (-1)^{-n} \Big[H(t - \lambda_1(\ell(2n+1) - y)) + H(t - \lambda_1(\ell(2n+1) + y)) - (H(t - \lambda_2(\ell(2n+1) - y)) - H(t - \lambda_2(\ell(2n+1) + y))) \Big] . \tag{7.26}$$

In this solutions clearly two waves with the wave velocities λ_i^{-1} are identified. With the same inverse transformations the time domain solutions of other boundary conditions are achieved.

General case of arbitrary κ. For an arbitrary value of κ a numerical inverse Laplace transformation is necessary. A number of methods are available in the literature, and the advantages and disadvantages has been studied, e.g., in [53] or [133]. But, in this case here, where one function in the convolution integral (7.24) is only available in Laplace and the other function in time domain, it is preferable to take the convolution quadrature method. This gives for the convolution integral (7.24)

$$u_y(n\Delta t) = \sum_{k=0}^{n} \omega_{n-k}(\Delta t) f(k\Delta t), \quad n = 0, 1, \dots, N, \tag{7.27}$$

with the weights $\omega_{n-k}(\Delta t)$ determined following formula (2.22)

$$\omega_{n-k}(\Delta t) = \frac{\mathscr{R}^{-(n-k)}}{L} \sum_{\ell=0}^{L-1} \hat{u}_y \left(\frac{\gamma\left(\mathscr{R} e^{i\ell\frac{2\pi}{L}}\right)}{\Delta t} \right) e^{-i(n-k)\ell\frac{2\pi}{L}} . \tag{7.28}$$

In the following, the time-dependent responses are evaluated with this method, choosing a backward differentiation formula of order 2 (BDF 2) as the underlying multistep method.

Poroviscoelastic solution. The poroviscoelastic solution of the one-dimensional problem given above is obtained by applying the elastic-viscoelastic correspondence principle to the solutions (7.13) - (7.18) [163]. In this case R and Biot's effective stress coefficient α are replaced by the corresponding expressions (6.15), respectively, with the complex moduli (6.16). Further, the modulus E used in the governing equations (7.1) and (7.2) is set to

$$E = K + \frac{4}{3}G \quad \Longrightarrow \quad \hat{E} = K\frac{1+q_k s^{\alpha^k}}{1+p_k s^{\alpha^k}} + \frac{4}{3}G\frac{1+q_g s^{\alpha^g}}{1+p_g s^{\alpha^g}} . \tag{7.29}$$

Inserting this complex modulus in either λ_i and d_i as well as in the solutions (7.13) - (7.18) itself gives finally the poroviscoelastic solutions in Laplace domain. Due to the complexity of the dependence to s no analytical inverse transformation was found. Therefore, as in the general poroelastic case, the convolution quadrature method is used to achieve a time-dependent solution for the poroviscoelastic 1-d column.

7.1.2 Poroelastic results

Wave propagation in the 1-d column sketched in Fig. 7.1 is studied in the following using the developed solutions. Three very different materials, a rock (Berea sandstone) [48], a soil (coarse sand) [114], and a sediment (mud) [14] are chosen to represent a wide range of porous materials. The material data are given in Table 7.1. In all calculations below it is assumed that the time history of the loading is a Heaviside step function.

Table 7.1. Material data of Berea sandstone, a soil, and a sediment (mud)

	K $\left(\frac{N}{m^2}\right)$	G $\left(\frac{N}{m^2}\right)$	$\rho\left(\frac{kg}{m^3}\right)$	ϕ	$K_s\left(\frac{N}{m^2}\right)$	$\rho_f\left(\frac{kg}{m^3}\right)$	$K_f\left(\frac{N}{m^2}\right)$	$\kappa\left(\frac{m^4}{Ns}\right)$
rock	$8\cdot10^9$	$6\cdot10^9$	2458	0.19	$3.6\cdot10^{10}$	1000	$3.3\cdot10^9$	$1.9\cdot10^{-10}$
soil	$2.1\cdot10^8$	$9.8\cdot10^7$	1884	0.48	$1.1\cdot10^{10}$	1000	$3.3\cdot10^9$	$3.55\cdot10^{-9}$
mud	$3.7\cdot10^7$	$2.2\cdot10^7$	1396	0.76	$3.6\cdot10^{10}$	1000	$2.3\cdot10^9$	$1\cdot10^{-8}$

First, to show the reliability of the proposed numerical algorithm (7.27), a comparison is made with Dubner and Abate's method [78]. For a finite column of length $\ell = 10$m subject to a stress only loading of $\sigma_y(t, y=\ell) = -1\,\text{N/m}^2$ and $p(t, y=\ell) = 0\,\text{N/m}^2$ at the top, the displacement $u_y(t, y=\ell)$ at the top and the pressure $p(t, y=0)$ at the bottom is plotted versus time in Fig. 7.2 and Fig. 7.3, respectively. For the convolution quadrature method, following suggestions concerning the choice of the parameters L and $\mathscr{R}$ as reported in Chap. 2, Δt is the only parameter to be adjusted. To test the convergence, several Δt values are chosen in the evaluation, with result plotted in dot and dash lines in Figs. 7.2 and 7.3.

Dubner and Abate's method, on the other hand, requires the empirical selection of two parameters, the real part of the Laplace variable s denoted as α, and the time period T. After a number of trial, the optimal values are chosen as $\alpha = 10$ for the

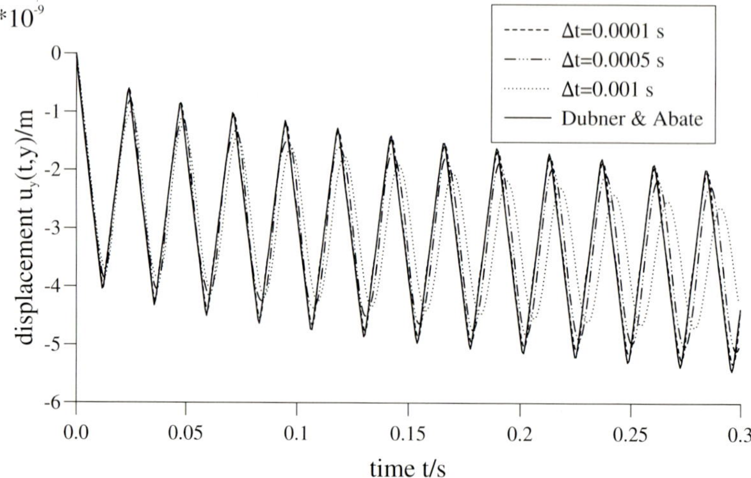

Fig. 7.2. Displacement $u_y(t, y = \ell)$ versus time for different time step sizes Δt compared with Dubner and Abate's inversion formula

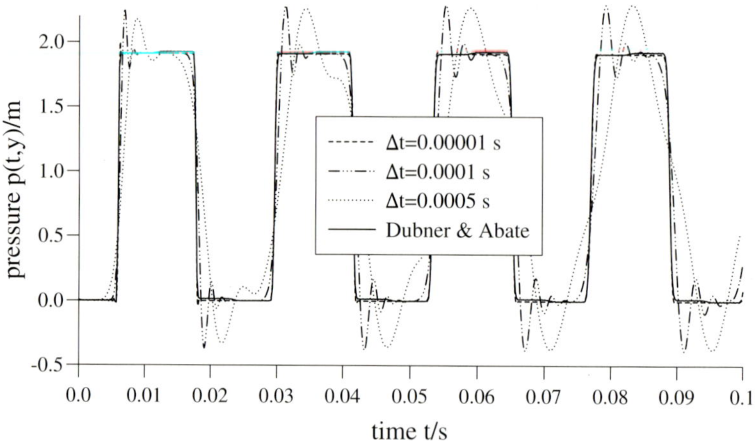

Fig. 7.3. Pore pressure $p(t, y = 0)$ versus time for different time step sizes Δt compared with Dubner and Abate's inversion formula

displacement and $\alpha = 30$ for the pressure solution and $T = 0.8 * t_{max}$, where t_{max} is the total observation time, for the soil case. The result is plotted in solid line in Figs. 7.2 and 7.3. For the two other cases, rock and sediment, multiple tries have failed to produce acceptable results. Hence only the soil case is compared.

In Figs. 7.2 and 7.3, it is observed that the accuracy of the convolution quadrature method is dependent on the time step size. If the time step Δt is small enough, the result overlaps with the Dubner and Abate result for most part of the curve. The convolution quadrature solution, however, shows a slight overshooting at the wave front of the pressure, whereas the Dubner and Abate method had a slight difficulty in keeping constant values between two wave fronts. This first test has shown the reliability and robustness of the proposed method.

Before moving to the next problem, it is of interest to provide a physical interpretation of the results observed in Fig. 7.2. For the purpose of comparison, it is first realized that for an elastic solution, the displacement at the top of column is given by triangular waves of constant amplitude that fluctuate around a constant mean value (see Sect. 4.3). In the poroelastic solution, the triangular wave form is largely preserved. The amplitude, however, diminishes with time due to fluid viscous dissipation, and will eventually go to zero. The mean value also drops with time. The mean fluctuation level is first around the static deformation value based on "undrained" material parameters, $u_{\text{static}} = 1.86 \cdot 10^{-9}\,\text{m}$. As sufficient fluid has gained time to escape at the top of column, a "consolidation" is observed. The mean fluctuation level gradually settles into the "drained" static deformation value $u_{\text{static}} = 2.94 \cdot 10^{-8}\,\text{m}$. As the soil column is being drained, the time for the wave to transverse the column will gradually increase.

Next, the pressure response in Fig. 7.3 is examined. The arrival time of the first compressional wave at the bottom of the column is clearly observed. The amplitude is twice of that created by static Skempton effect due to the perfect reflection condition at the bottom. From the well known one-dimensional wave propagation in a fixed-free end column, square waves are expected. If enough number of cycles are observed, the waves will eventually drop to zero due to dissipation.

In the above observed time range the second compressional wave, known as the slow wave is not detected. This is attributed to the large ratio in wave speed such that the fast wave has the opportunity to transverse the column a number of times before the arrival of the slow wave at the bottom.

To unambiguously capture the slow wave, next, an "infinite" column is considered to avoid wave reflections. This is achieved by using a column length of $\ell = 1000\,\text{m}$ and a short observation time. In Fig. 7.4, the pressure, $p\,(t, y = 995\,\text{m})$, five meters behind the excitation point ($y = \ell = 1000\,\text{m}$) is depicted versus time. Since this is the first time that such wave can be observed, it is compared with the exact time domain solution (7.26), shown as solid lines in Fig. 7.4 for the three materials, to gain confidence. To make the comparison, an arbitrarily large value, $\kappa = 1 \cdot 10^{-2}\,\text{m}^4 / (\text{Ns})$, is chosen in the convolution quadrature solution, with results plotted in dashed lines in Fig. 7.4. It is observed that, except for some fluctuations at wave fronts, which are generally unavoidable for all numerical inversion methods, the two solutions compare very well.

The phenomenon exhibited in Fig. 7.4 can be rationalized as follows. First, the fast wave arrives at $y = 5\,\text{m}$ causing the step jump. The second wave, arriving at a later time, is of negative amplitude and cancels exactly the first wave as indicated by the exact solution (7.26). The arrival time of the two waves is independent of κ as its limit has been taken.

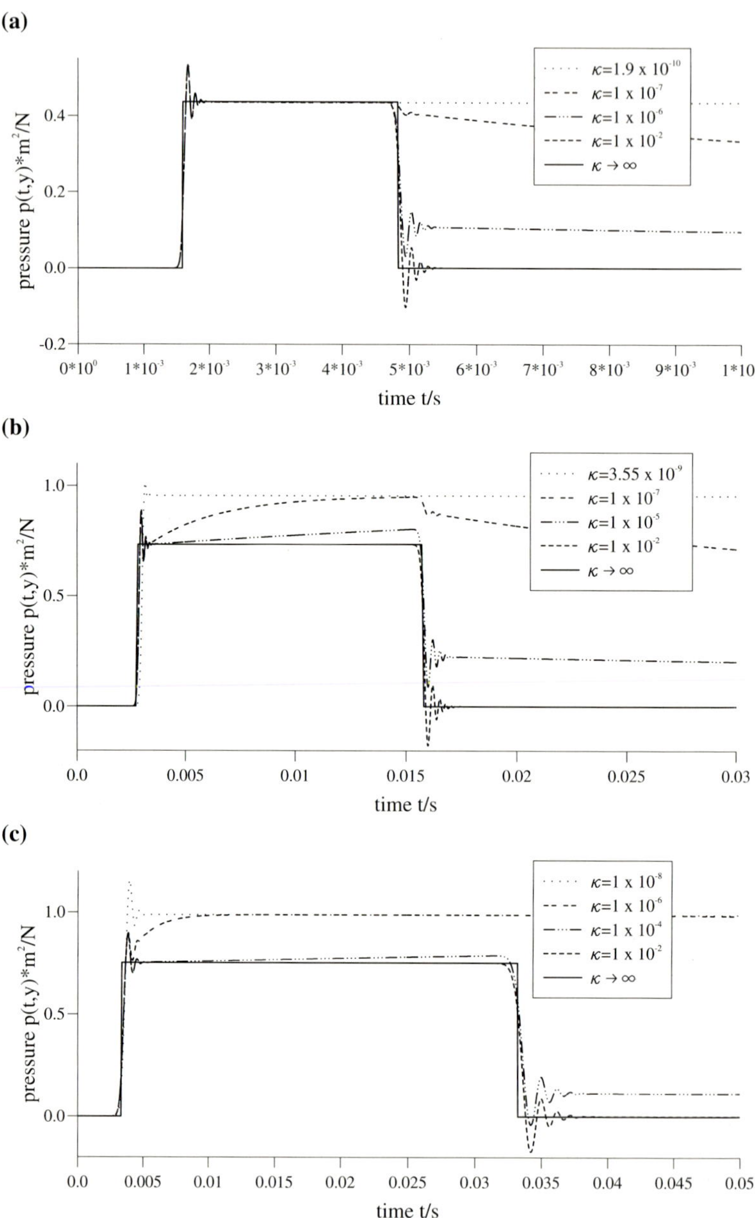

Fig. 7.4. Pressure $p(t, y = 995\,\mathrm{m})$ versus time for different values of κ compared with the analytical solution (7.26): Infinite column **(a)** Berea sandstone **(b)** soil **(c)** sediment

To obtain and understand the solution of the realistic cases, κ values are decreased. Figure 7.4 shows a sequence of reduction that lead to the real values listed in Table 7.1. As κ decreases, it is observed that both the amplitude and the arrival time of the waves are affected. The effect is strongest for the second wave. For some intermediate values of κ the amplitude is diminished, when the second wave arrives. Hence the pressure does not drop to zero at the passage of the wave front. Also the second wave is identified to be dispersive as it does not arrive as a sharp front with constant value in some cases. Rather, the pressure continues to decline as seen in some curves.

As κ continue to decrease, two effects appear. First, the wave speed of the second wave tends to zero as $\kappa \to 0$. Second, the wave is rapidly dissipated such that it has no effect when it arrives at the 5 m point. In that case, only the arrival of the first wave is observed, and not the second wave. These observations are in accordance with the behavior of λ_i with increasing and decreasing κ.

Also, the comparison shows the different behavior of the three different materials on changing the permeability. For rock, the wave amplitude of the first wave is nearly independent from the permeability, contrary to the soil and the sediment.

If the same experiment is examined with a finite soil column, now $\ell = 10\,\mathrm{m}$, the reflections at both ends are visible, and there are multiple arrivals (see Fig. 7.5). It is

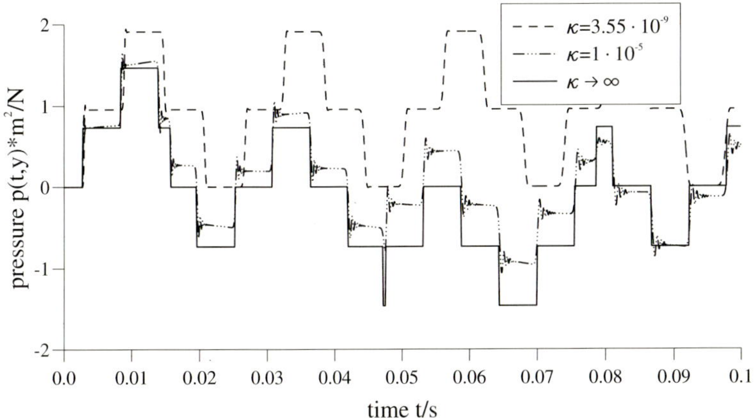

Fig. 7.5. Pressure $p(t, y = 5\,\mathrm{m})$ versus time for different values of κ compared with the analytical solution: Finite soil column

of interest to check the successive arrival time with theoretical result. However, the wave speed is not a constant when there exists dissipation. Only an estimate based on the special case of $\kappa \to \infty$ can be used. In Table 7.2, wave speed for the special case is presented as λ_i^{-1}. The successive arrival times in the middle of the column (5 m) are shown as 1st, 2nd, etc. With these values, results for the undamped case can be interpreted. Referring to Fig. 7.5, the pressure for the undamped case at 5 m is shown in solid line. At $t = 0.0028\,\mathrm{s}$, the arrival of the fast wave is observed. At

Table 7.2. Arrival times of the two waves at $y = 5\,\mathrm{m}$ in the finite column: Material data of soil

	λ_i^{-1}	1st	2nd	3rd	4th	5th	6th
faster wave	$1788\,\mathrm{m/s}$	0.0028 s	0.0084 s	0.0140 s	0.0200 s	0.0252 s	0.0308 s
slow wave	$318\,\mathrm{m/s}$	0.0157 s	0.0471 s	0.0786 s	0.1100 s	0.1415 s	0.1730 s

$t = 0.0084\,\mathrm{s}$, the bottom reflected fast wave arrives. Next comes the top reflected fast wave at $t = 0.0140\,\mathrm{s}$. At $t = 0.0157\,\mathrm{s}$, the arrival of the slow wave negates the pressure. This identifying process can be continued for every arrival front.

The more interesting case is the real case with dissipation. Two κ values are used. For the intermediate value case, $\kappa = 1 \cdot 10^{-5}\,\mathrm{m}^4/(\mathrm{Ns})$, significant modification of wave amplitude, especially after multiple reflections, are observed. The arrival time is roughly the same as the undamped case. For the smallest permeability (actual value) case, the effect of slow wave is not visible. The wave profile is similar to the elastic case. However, a closely tracking of the arrival time and comparison with the undamped case indicates that the wave slows down after each reflection. This behavior is in accordance with the theory where λ_i are functions of s, hence are time-dependent, leading to time-dependent wave velocities. Additionally, the sharp wave fronts are smoothed after each reflection, and the wave amplitude diminishes with time.

Finally, the wave propagation with respect to both temporal and spatial variables is considered. In Fig. 7.6, the displacement $u_y(t,y)$ caused by a stress Heaviside step loading is depicted versus time and at the locations $y = 2.5\,\mathrm{m}, 5\,\mathrm{m}, 7.5\,\mathrm{m}, 10\,\mathrm{m}$. In this figure, the Berea sandstone data are used with two different permeabilities to show the extreme case of vanishing damping compared to the realistic damping. The realistic case is dominated by the first compressional wave, as expected from the previous study. In the undamped case the faster wave is a kind of overtone to the slower wave.

In the next case, the boundary condition is changed to a pressure Heaviside step loading of $1\,\mathrm{N/m}^2$, while the total stress is zero. Although this case is physically unattainable, it is mathematically valid, and the result can be used in a superposition. If the top of column is exposed to a fluid and a step pressure rise is applied, the boundary condition consists of a stress part, and a pressure part, of which the current solution represents. This case is presented to bring the relation of a pressure loading and the second wave into sharp focus.

In Fig. 7.7, the pressure $p(t,y)$ versus time at the locations $y = 0\,\mathrm{m}, 2.5\,\mathrm{m}, 5\,\mathrm{m}, 7.5\,\mathrm{m}$ is presented for Berea sandstone and the same permeabilities as before. The influence of κ is much stronger for pressure loading than for stress loading. In part (a), the small permeability case, the maximum amplitude is much smaller than that in part (b), and the wave propagates much faster. If part (b) is plotted in logarithmic scale, waves of very small amplitudes leading the large wave front shown in the figure could be observed. Hence the wave front observed in (b) is the slow wave with wave speed $\lambda_1^{-1} = 1037\,\mathrm{m/s}$. The first wave is not seen because it is too small.

(a)

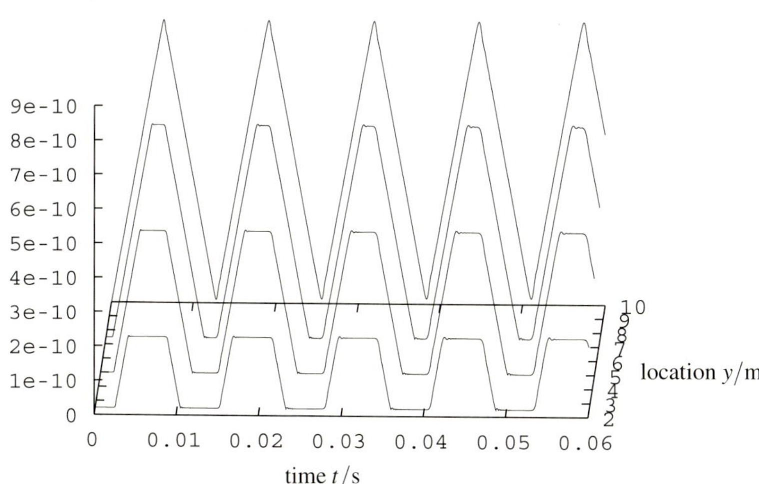

(b)

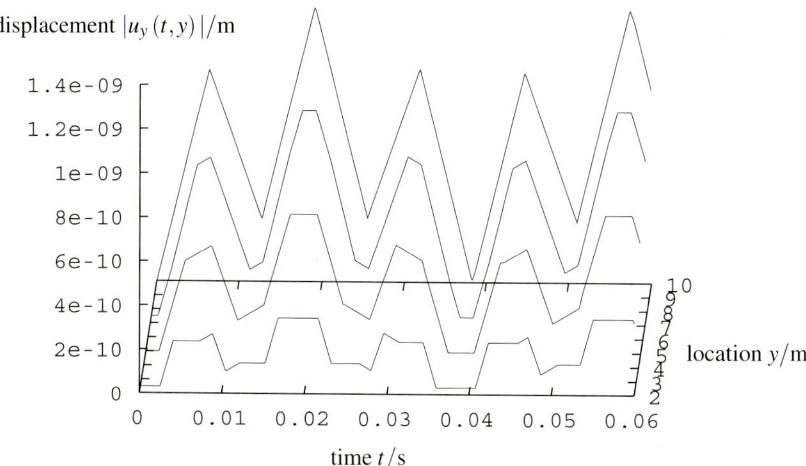

Fig. 7.6. Displacement $|u_y(t,y)|$ (absolute value) versus time at different locations y: Finite rock column **(a)** $\kappa = 1.9 \cdot 10^{-10}$ **(b)** $\kappa \to \infty$

(a)

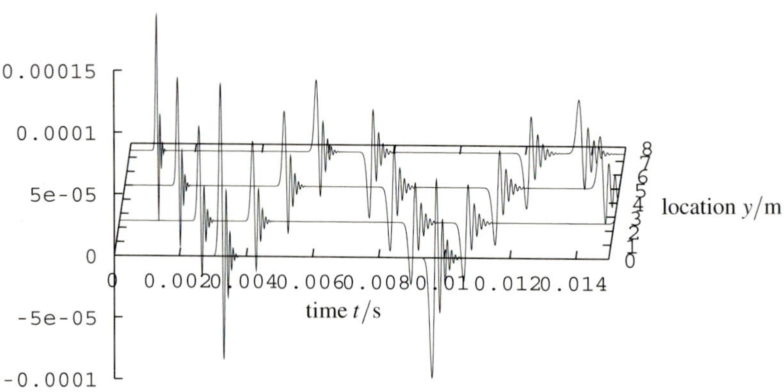

(b)

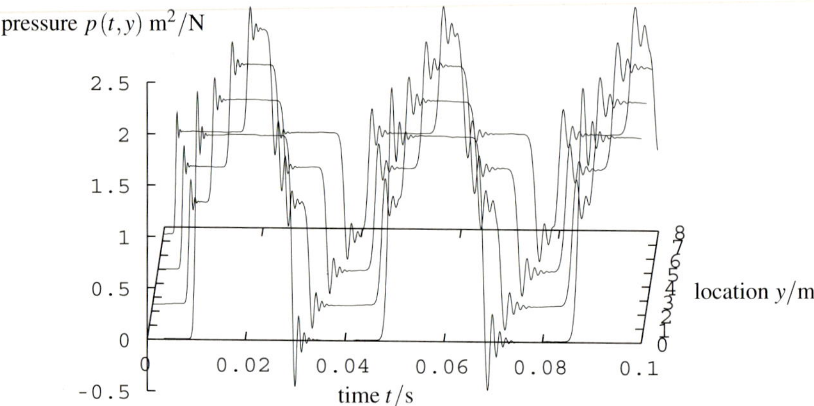

Fig. 7.7. Pressure $p(t,y)$ versus time at different locations y: Finite rock column **(a)** $\kappa = 1.9 \cdot 10^{-10}$ **(b)** $\kappa \to \infty$

It is recognized that the pressure boundary condition generates a second wave that travels undamped due to the high permeability condition. While the top of the column is total stress free, there is no compression generated at that point. A first wave is not generated at the boundary. As the second wave travels through the column, it emits the first wave of small amplitude, which outruns the second wave.

For the top figure (a), the small permeability case, the first wave is recognized by checking the approximate wave speed $\lambda_2^{-1} = 3137\,\mathrm{m/s}$. The first wave is observed only by plotting in the scale shown. We can use such small scale because unlike case (b), the second wave is all but vanished in amplitude when it reaches the observation points. Although the second wave survived only a short distance, the first wave that it generated is observed in this figure. The first wave does not have a sharp front because it is continuously emitted by the second wave.

7.1.3 Poroviscoelastic results

In the studies above an elastic skeleton was assumed. Next, viscoelastic effects will be studied as introduced in the constitutive equations (6.13) and (6.14). In these constitutive equation the bulk modulus $\hat{K}$, the shear modulus $\hat{G}$, and the compression modulus of the solid itself $\hat{K}_s$ are each chosen to be viscoelastic, modeled by a three-parameter model. For each of them, the values of p and q need to be given. However, to the author's best knowledge, no such data have been reported in the literature. Therefore, the same set of data is somewhat arbitrarily chosen for the three materials. To compare the influence of viscoelasticity in different moduli on the dynamic response, four different cases are considered

Case 1: Only the bulk compression modulus $\hat{K}(s)$ is modeled viscoelastic: $p_k = 1\,\mathrm{s}^{-1}$, $q_k = 1.5\,\mathrm{s}^{-1}$ and $p_{ks} = p_g = q_{ks} = q_g = 0\,\mathrm{s}^{-1}$

Case 2: Only the shear modulus $\hat{G}(s)$ is modeled viscoelastic: $p_g = 1\,\mathrm{s}^{-1}$, $q_g = 1.5\,\mathrm{s}^{-1}$ and $p_{ks} = p_k = q_{ks} = q_k = 0\,\mathrm{s}^{-1}$

Case 3: Only the compression modulus of the solid material $\hat{K}_s(s)$ is modeled viscoelastic: $p_{ks} = 1\,\mathrm{s}^{-1}$, $q_{ks} = 1.5\,\mathrm{s}^{-1}$ and $p_k = p_g = q_k = q_g = 0\,\mathrm{s}^{-1}$

Case 4: The purely poroelastic case without any viscoelasticity: $p_{ks} = q_{ks} = p_k = p_g = q_k = q_g = 0\,\mathrm{s}^{-1}$

Before solving the transient problems, the frequency response of a column with length $\ell = 1\,\mathrm{m}$ is considered. In Fig. 7.8, the absolute value of the displacement $\hat{u}_y(\omega, y = \ell)$ at the top of the column is plotted versus frequency ω for the three materials. As boundary condition, a constant step pressure loading (without total stress) is assumed. In Fig. 7.8, the expected resonance peaks are found. The first resonance frequency is around 2000 Hz for the sediment, which increases to about 5000 Hz for the rock. The various curves correspond to different assumptions of viscoelasticity, referred to as case 1 to 4 in the above. It is found that the sediment response is least affected by viscoelastic effect–there is basically no shift in eigenfrequencies and only a slight damping in response amplitude. This is in accordance with our model, because the sediment bulk property is dominated by the fluid, which is elastic. The

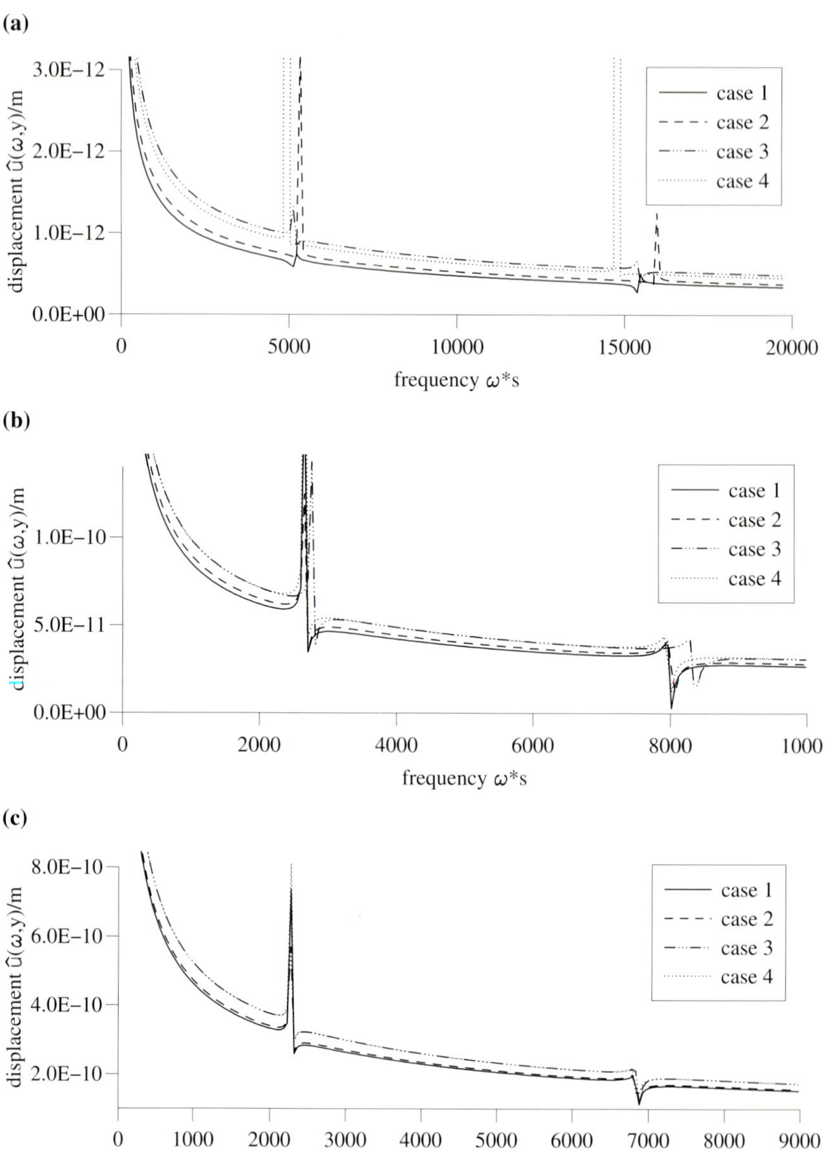

Fig. 7.8. Absolute value of the displacement $|\hat{u}_y(\omega, y = \ell)|$ at the top of the column versus frequency ω **(a)** Berea sandstone **(b)** soil **(c)** sediment

viscoelastic solid hence contributes to only a secondary influence. The soil response is also less influenced. There is a small shifting of eigenfrequencies, and a somewhat larger damping than the sediment case. The largest effects are found in the rock material. Not only there exists larger damping, particularly on the resonance peaks, but also significant shift of eigenfrequencies occurs. Further, it is noted that for all materials, the largest damping results from the viscoelasticity of bulk compression modulus. For soil, the largest shift of eigenfrequencies results from the viscoelastic effect of $\hat{K}_s$, compared to rock where $\hat{G}$ has the most influence. This shows that the effect of each modulus is different in different materials.

For the frequency response of the other two boundary conditions, a stress and a displacement loading, the influence of viscoelasticity exhibits similar trend. Hence it is enough to show the results for just this boundary condition.

Now, the time-dependent behavior is considered. Due to the relative insensitivity of sediment response to viscoelasticity, only results for the two other materials are presented. In Fig. 7.9, the displacement $u(t, y = \ell)$ at the top of the column, caused by a step stress loading $\sigma(t, y = \ell) = -1\,\mathrm{N/m^2}\,H(t)$, is depicted versus time. In each of the curves, a different time step size is used for the convolution quadrature method, due to different wave speeds of the materials. For the Berea sandstone $\Delta t = 1 \cdot 10^{-5}\,\mathrm{s}$ and for the soil $\Delta t = 2 \cdot 10^{-5}\,\mathrm{s}$ are used, with $N = 500$ time steps.

In Fig. 7.9, the rock displacement shows an oscillation similar to that for an elastic material, whereas for the soil, the oscillation is combined with a settlement, due to the well-known consolidation effect. It is noticed that the wave speed is modified in both materials. Case 4, the case without viscoelasticity, has the slowest wave speed, by observing the time it takes the wave to transverse the column. This is not surprising, because by setting the two parameters p and q of the three-parameter model constitutive equation (see Fig. 5.1) to zero, case 4 has the smallest modulus. In the viscoelastic cases, the apparent modulus of the material is between $1.5E$ for small time (or high frequency), and E for large time (or low frequency), due to the p and q values used. Hence, the wave speed of the viscoelastic and the elastic cases should not be directly compared. However, among the viscoelastic cases, it can be compared and observed that different modulus has different effect on the two materials. The fastest wave in the rock is associated with the viscoelasticity of shear modulus. The fastest wave in soil, on the other hand, is observed to be associated with the solid compression modulus. The oscillation amplitude is found to be the smallest also in these two cases, respectively for soil and rock. These are consistent with the observation in frequency domain.

Next, wave propagation in an "infinite" ($\ell = 1000\,\mathrm{m}$) 1-d column is investigated with the aim of capturing the two compressional waves, a fast and a slow wave, as also done in the case with the elastic skeleton. The results concerning the behavior with respect to different values of κ are not influenced by the viscoelasticity. Therefore, only the influence of viscoelasticity in the individual modulus is studied in Fig. 7.10. As in Fig. 7.4, the pressure $p(t, y = 995\,\mathrm{m})$ due to a stress Heaviside step loading is plotted versus time. But, here, the cases 1 to 4 defined in the beginning of this section are examined. To enhance the observation of the second wave, a large

(a)

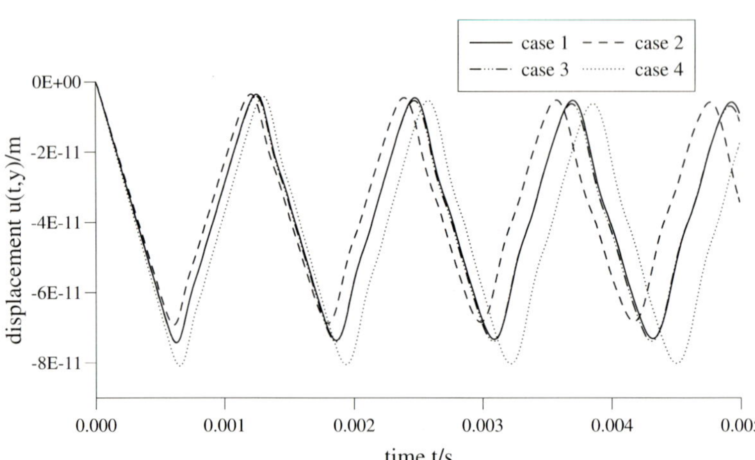

(b)

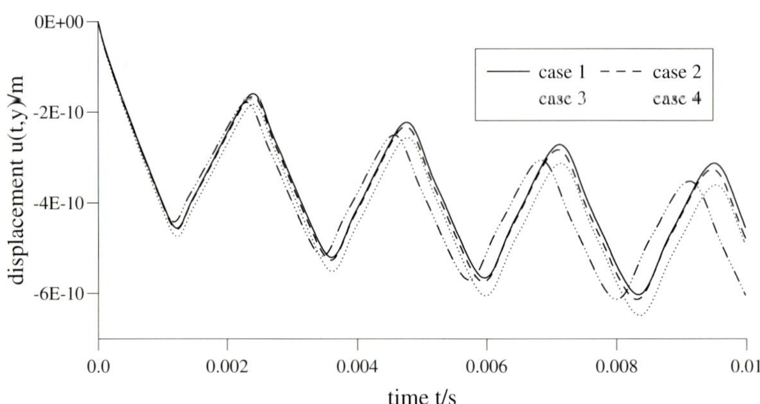

Fig. 7.9. Displacement $u_y\,(t, y=\ell)$ at the top of the column versus time t **(a)** Berea sandstone **(b)** soil

permeability, $\kappa = 10^{-2}\,\mathrm{m}^4/(\mathrm{Ns})$, is used here. Similar to the investigation above, the viscoelasticity of different modulus has different effects on the two materials. First of all, it is observed that the wave velocities are modified, much more so for the second wave than for the first wave. The arrivals of the first waves are close to each other. Nevertheless, in both materials case 4 gives the slowest first wave. In rock, case 2 has the fastest first wave, and in soil, it is case 3. These are consistent with earlier observations. The second wave, on the other hand, is more complicated. In most cases the second wave of the viscoelastic cases travels faster than the non-viscoelastic one, case 4. However, in case 3, where only the solid grain modulus is modeled viscoelastic, the first wave becomes faster, but the second wave becomes

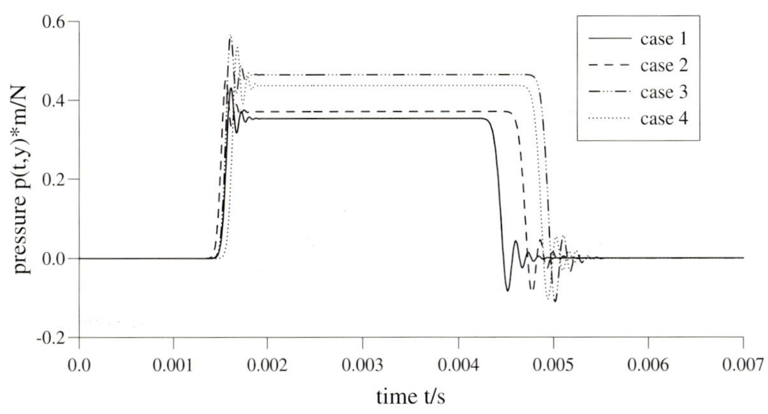

(b)

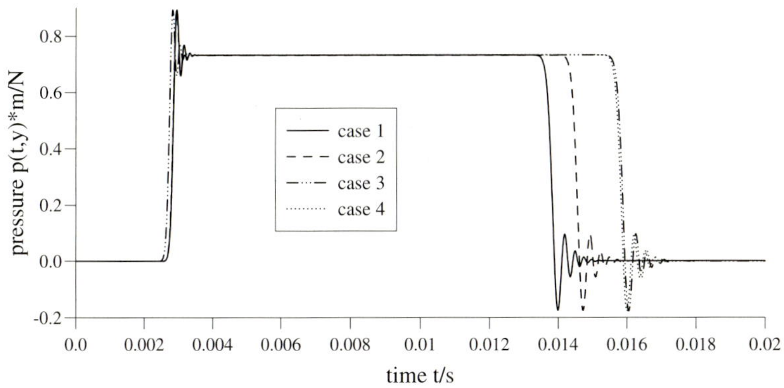

Fig. 7.10. Pressure $p(t, y = 995\,\mathrm{m})$ versus time: Wave propagation for different damping cases **(a)** Berea sandstone **(b)** soil

slower than case 4. It is also observed that there is significant amplitude reduction of the first wave for the rock material when viscoelasticity is present, except for case 3, where the amplitude increases. For the soil, there is little change in amplitude.

Summarizing, the presented results either poro- or poroviscoelastic clearly show two waves and their interplay under different loading and boundary conditions. But, for realistic values of permeability of the three used materials, the influence of the second compressional wave is small. Changing the permeability by using a different fluid, such as air, with low viscosity, especially in the case of pressure boundary conditions, a stronger influence of the second wave may be observed. Further, it is shown that viscoelastic effect is stronger in rock and soil, than in sediment. The rock is shown to be more influenced by the shear modulus whilst the soil is more affected by the compression modulus of the grains. The conclusions drawn here concerning the viscoelastic effect are not entirely general, and are material dependent.

7.2 Waves in half space

The propagation of waves in a 3-d half space is studied by the presented time-stepping boundary element formulation for poro- and viscoelastic media. The underlying multistep method $\gamma(z)$ is a BDF 2 and $L = N$ is chosen as suggested in [161].

7.2.1 Rayleigh surface wave

Dealing with wave propagation in a half space, surface waves are one of the most interesting effects. Especially, the Rayleigh wave is of interest due to its disastrous consequence in earthquakes. This surface wave caused by wave reflections at the free surface was first investigated by Lord Rayleigh [147], who has shown that its effect decreases rapidly with depth and its velocity of propagation is smaller than that of a body wave. This wave velocity can be approximated by the formula [98]

$$c_R = \frac{0.87 + 1.12\nu}{1+\nu} c_2 \,. \tag{7.30}$$

Analytically, the Rayleigh wave is found in the solution presented by Pekeris [143]. He assumed a point load on the traction free surface of an elastic half space. The load has a Heaviside time history, i.e., starts acting at time $t = 0\,\mathrm{s}$ and is then kept constant. However, the elastic material parameters can not be chosen arbitrarily in this solution, Poisson's ratio is fixed on $\nu = 0.25$, whereas the Young's modulus is free. In Fig. 7.11, Pekeris analytical solution for the vertical displacement at $15\,\mathrm{m}$ distance from the excitation point is presented assuming $E = 2.5 \cdot 10^8\,\mathrm{N/m^2}$. There, the arrival of the fast compression wave ($t = 0.037\,\mathrm{s}$) is identified as the first deviation from zero value. Contrary, the arrival of the shear wave ($t = 0.065\,\mathrm{s}$) is not visible due to the strong increase of the displacement value up to infinity. This pole, sometimes called Rayleigh pole, indicates the arrival of the Rayleigh wave ($t = 0.70\,\mathrm{s}$).

To capture this pole for arbitrary elastic material data, i.e., Poisson's ratio $\nu \neq 0.25$, for a viscoelastic, or a poroelastic half space the proposed boundary element formulation will be used. Because in the boundary element formulation applied here a full space fundamental solution is used, the free surface has to be discretized. A long strip ($6\,\mathrm{m} \times 30\,\mathrm{m}$) is discretized with 396 triangular linear elements on 242 nodes (see Fig. 7.12). The used time step size is $\Delta t = 0.0006\,\mathrm{s}$. The modeled half space is loaded on area A $(1\,\mathrm{m}^2)$ by a vertical total stress vector $t_z = -1000\,\mathrm{N/m^2}\,H(t)$ (shaded area in Fig. 7.12) and the remaining surface is traction free. In case of modeling the half space material poroelastic, the pore pressure is assumed to be zero all over the surface, i.e., the surface is permeable.

The material properties are those of the soil used in Sect. 7.1. In the following, the wave propagation is studied not only for a poroelastic modeled soil but also for

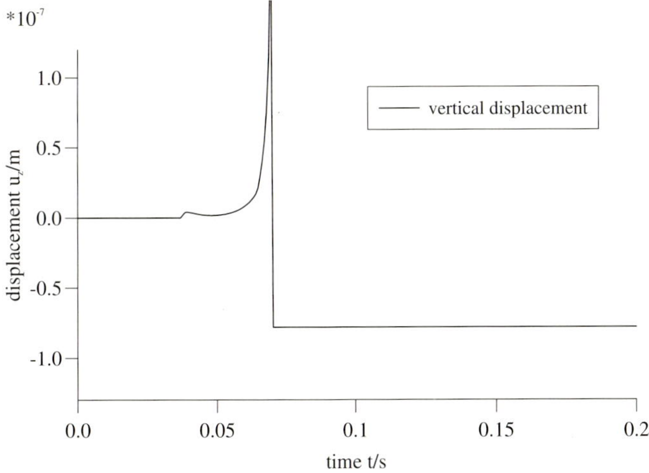

Fig. 7.11. Vertical displacement of a elastic half space at 15 m distance from excitation point: Pekeris analytical solution

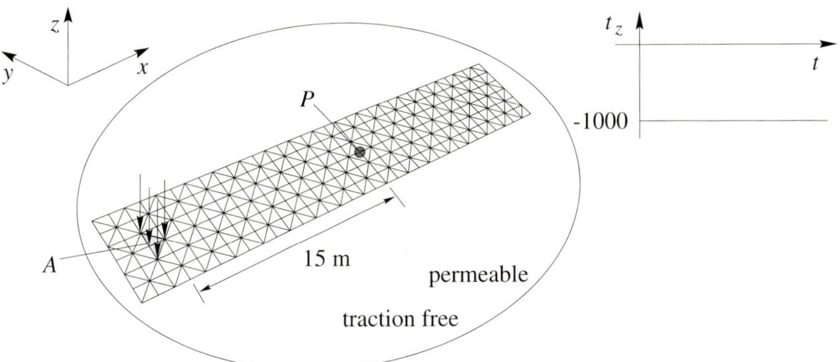

Fig. 7.12. Half space under vertical load: Discretization and load history

an elastic or viscoelastic modeling. The two elastic material models, drained and undrained, used, are the both cases from Sect. 6.4.2 using the same shear modulus as the poroelastic material but either drained Poisson's ratio or undrained Poisson's ratio (see Table 7.3). As discussed in Sect. 6.4.2, these two elastic models should be an upper and lower bound for the poroelastic material. Since no measured data for a viscoelastic material model of the soil are available, the elastic constants E, ν and ρ are taken form the drained elastic material, and the damping coefficients p^D, q^D, and α^D are chosen arbitrarily. It is assumed that only the deviatoric part of the stress-strain relation is viscoelastic, whereas the hydrostatic part is elastic. The following results will reflect this assumption. The data for all three material models are summarized in Table 7.3.

Table 7.3. Material data of a soil (coarse sand) modeled poroelastic, elastic, and viscoelastic

poroelastic							
$K\left(\frac{N}{m^2}\right)$	ϕ	$G\left(\frac{N}{m^2}\right)$	$\rho\left(\frac{kg}{m^3}\right)$	$K_s\left(\frac{N}{m^2}\right)$	$\rho_f\left(\frac{kg}{m^3}\right)$	$K_f\left(\frac{N}{m^2}\right)$	$\kappa\left(\frac{m^4}{Ns}\right)$
$2.1\cdot10^8$	0.48	$9.8\cdot10^7$	1884	$1.1\cdot10^{10}$	1000	$3.3\cdot10^9$	$3.55\cdot10^{-9}$

elastic					
drained			undrained		
$E\left(\frac{N}{m^2}\right)$	ν	$\rho\left(\frac{kg}{m^3}\right)$	$E\left(\frac{N}{m^2}\right)$	ν	$\rho\left(\frac{kg}{m^3}\right)$
$2.5\cdot10^8$	0.298	1884	$2.9\cdot10^8$	0.49	1884

viscoelastic								
$E\left(\frac{N}{m^2}\right)$	ν	$\rho\left(\frac{kg}{m^3}\right)$	$p^H\,(s^{-1})$	$q^H\,(s^{-1})$	α^H	$p^D\,(s^{-1})$	$q^D\,(s^{-1})$	α^D
$2.5\cdot10^8$	0.298	1884	0	0	1	0.8	1	1

The following results can better be understood when the wave velocities of the above materials are known to identify the arrival time of the different waves at point P (see the mesh in Fig. 7.12). In Table 7.4, the compression wave velocity c_1, the shear wave velocity c_2, and the Rayleigh wave velocity c_R are given together with their arrival time, where the Rayleigh wave velocity is approximately determined by equation (7.30). Because the wave velocities of the viscoelastic material are known

Table 7.4. Wave velocities and corresponding arrival times for the soil

	c_1	c_2	c_R	t_1	t_2	t_R
drained	425 m/s	228 m/s	211 m/s	0.035 s	0.066 s	0.071 s
undrained	1629 m/s	228 m/s	217 m/s	0.009 s	0.066 s	0.069 s
viscoelastic	445 m/s	255 m/s	236 m/s	0.034 s	0.059 s	0.063 s

to be time-dependent, in Table 7.4 the initial values calculated with formula (B.5) are given. Further, the Rayleigh wave velocity is calculated with formula (7.30) assuming that this formula deduced for elasticity is also a good approximation in case of viscoelasticity. The real arrival times of the viscoelastic waves can be larger due to dissipation. For the poroelastic modeled soil, no wave velocities are given due to their strong time dependence. Also, no simple approximation for them as in the case of viscoelasticity is available [14].

First, the surface displacement at point P in 15 m distance from the excitation point is presented in Fig. 7.13 for the drained elastic and the viscoelastic modeled soil. The arrival time of the compression wave ($t \approx 0.035$ s) is identified for both the vertical and horizontal displacement solution. In both solutions, the arrival time for the two material models is nearly the same, whereas the pre-calculated values differ.

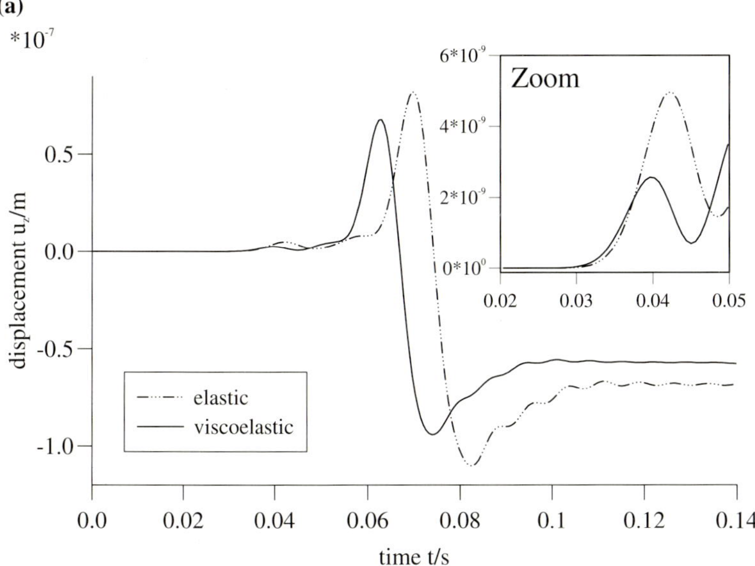

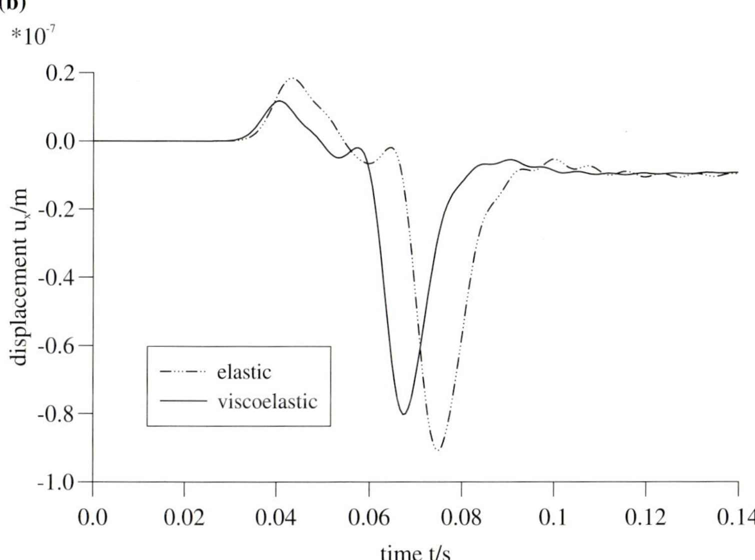

Fig. 7.13. Vertical and horizontal displacement at point P versus time: Comparison viscoelastic and elastodynamic modeling of the soil **(a)** vertical **(b)** horizontal

But, this difference is less than two time steps, and, therefore, it can not be expected to identify this difference on the plot.

The next wave which arrives at point P is the shear wave followed closely by the Rayleigh wave. However, only the arrival of the Rayleigh wave is observed identified in the vertical displacement component as a positive pole and in the horizontal component as a negative pole. This in accordance with the behavior of the analytical Pekeris solution (see Fig. 7.11). Similar to the analytical solution, in the numerical solution in Fig. 7.13 the shear wave front can not be distinguished from the Rayleigh pole. The strong gradient of the displacement values before the arrival of the Rayleigh wave makes it impossible to detect there the shear wave front.

For times $t > 0.1\,\mathrm{s}$, the elastic and the viscoelastic solution tend to constant static values. The viscoelastic horizontal displacement component even reach the same static value as the elastic solution, whereas the vertical displacement component of the viscoelastic solution has a smaller absolute value as the elastic solution. However, the calculated results for theses large times are only qualitatively correct, i.e., the elastic solution only tends to the correct static solution. More quantitative correct results need an enlarged discretized area on the half space surface [5].

Summarizing the observations in Fig. 7.13, the displacement values are smaller and the shear wave velocity and the Rayleigh wave velocity of the viscoelastic medium is increased compared to the corresponding elastic values. This confirms the results achieved in Sect. 5.4: the viscoelastic model of the soil leads, basically, to a more stiff half space than an elastic modeling. But, the compression wave velocity seems to be the same for both material models. This is, presumably, caused by the assumption that only the deviatoric part of the stress-strain relation is viscoelastic contrary to the hydrostatic part which is assumed to be elastic.

Next, in Fig. 7.14, the time history of the vertical and horizontal displacement component at point P is depicted versus time for the poroelastic soil and both elastic modeled soils. The general behavior is the same as before for the viscoelastic modeled soil. Though, all comments concerning the time history of the displacements and the arrival times of the waves made for Fig. 7.13 can be transfered to Fig. 7.14.

Hence, the discussion of the results will concentrate on the poroelastic media. Since waves in poroelastic media are dispersive, i.e., the wave velocities are time-dependent, the waves are already damped when arriving at point P. This may be the reason for the modest increase of the displacement components u_z and u_x at the compression wave arrival ($t \approx 0.01\,\mathrm{s}$). The arrival time is that of the compression wave of the undrained elastic medium which coincides with the value given in Table 7.4. The compression wave of the drained elastic medium arrives later ($t \approx 0.035\,\mathrm{s}$). The shear wave of both elastic media arrive at the same time since both materials have the same shear modulus. The dissipation of the poroelastic material is expressed not only in the modest increase at the wave front but also in the small absolute displacement values until the shear wave arrives. Even before this arrival ($0.05\,\mathrm{s} < t < 0.06\,\mathrm{s}$), negative values are visible contrary to the elastic medium where the negative displacement values are found at $t > 0.06\,\mathrm{s}$. Summarizing, the

(a)

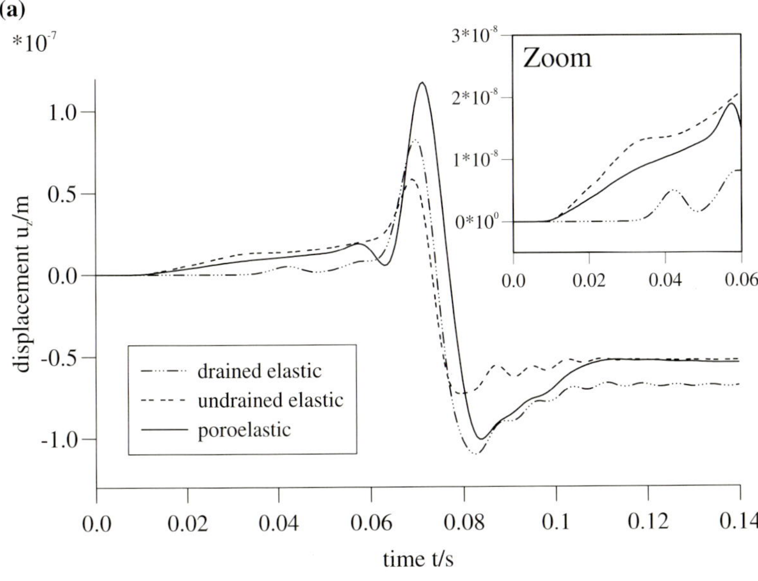

(b)

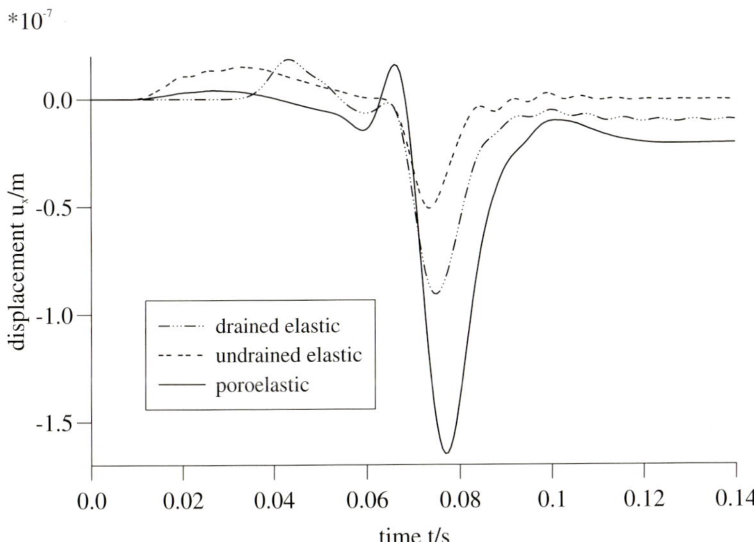

Fig. 7.14. Vertical and horizontal displacement at point P versus time: Comparison poroelastic and elastodynamic modeling of the soil **(a)** vertical **(b)** horizontal

compression wave in the poroelastic medium has less influence as in an elastic modeled medium.

Contrary, the Rayleigh pole in the vertical component and in the horizontal component of the poroelastic displacement solution is more pronounced than in the elastic ones. The displacement values are nearly twice that of the elastic values. After the arrival of the Rayleigh wave ($t > 0.09\,\mathrm{s}$), the vertical poroelastic displacement component lies in between the elastic solutions, i.e., they are an upper and a lower bound, whereas the horizontal poroelastic component is larger than both corresponding elastic values, i.e., for the horizontal component the two elastic cases are no upper and lower bound. These results and the comparison with the viscoelastic results give reason for the conclusion that modeling a fluid saturated soil as a one-phase material is only a very crude approximation of the real behavior.

As remarked at the beginning of this section, the effect of the Rayleigh wave decreases with depth. To visualize this physical property, the displacements at several points below the surface are considered. The observation points are put on a circle with radius of 15 m measured from the excitation point in area A. This ensures that the body waves, i.e., the compression wave and the shear wave, arrive at all points at the same time. The geometry is shown in the upper left of Fig. 7.15. In the same

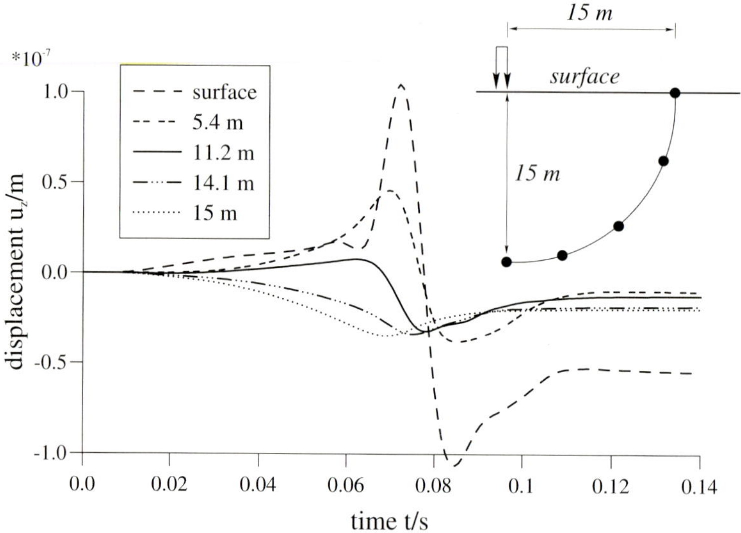

Fig. 7.15. Vertical displacement at points below the surface: Decrease of the effects of the Rayleigh wave with depth

figure, the time history of the vertical displacement and in Fig. 7.16 the pressure is presented.

First, the displacement solution (Fig. 7.15) is discussed. There, the vanishing influence of the Rayleigh wave with depth is observed. Mostly, at 14 m no longer any effects are visible, whereas until 5.4 m the Rayleigh pole is found. For larger times

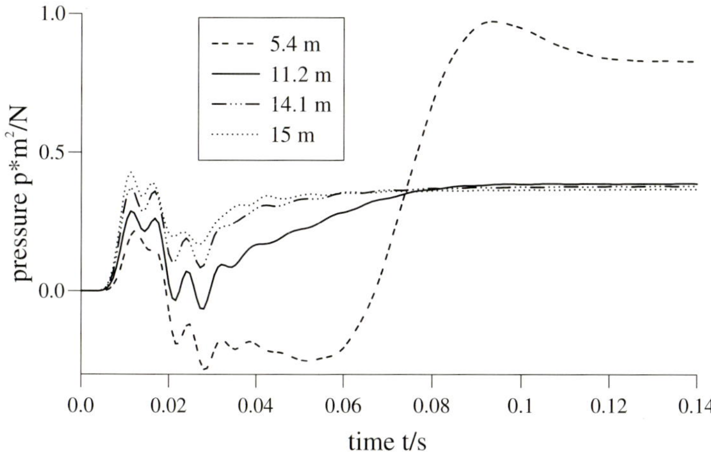

Fig. 7.16. Pore pressure at points below the surface: Decrease of the effects of the Rayleigh wave with depth

$t > 0.1$ s, the displacement amplitudes vary with depth only slightly except that the surface solution is much larger than the other values. This is partly caused by the Rayleigh wave but also and probably much stronger influenced by the geometrical damping.

The pressure solution (Fig. 7.16) has an interesting effect. First, at $t \approx 0.01$ s the compression wave produces a pressure jump with a subsequent creeping to the final value. This description is not valid for a depth of 5.4 m. There, after the jump the pressure solution has negative values. This is caused by the effect described in Sect. 6.4.2. At this time, a lifting of the surface due to the compression wave causes negative pressure in the surrounding because water is sucked in. But, this is only a local effect and for larger depth not visible.

As the fluid itself is modeled without shear components there exists no shear wave, and, consequently, no Rayleigh wave. However, the pressure solution has a jump at $t \approx 0.07$ s in the depth of 5.4 m. This must be an effect of the Rayleigh wave because this effect diminishes with depth and is exactly at the arrival time of the Rayleigh wave. This shows that the Rayleigh wave in the solid skeleton induces a compression wave in the fluid.

7.2.2 Slow compressional wave in poroelastic half space

One of the main differences between wave propagation in an elastic and a poroelastic solid is the second compressional wave, the so-called slow compressional wave. Theoretically, this wave has been found by Biot [31, 32]. In experiment, this wave type has been identified by Plona in [144]. Additionally, this slow wave can be confirmed by the 1-d solution mentioned above [164]. Here, now, this slow wave will be captured in a poroelastic half space loaded by total stress $t_z = -1\,\mathrm{N/m^2}\,H(t)$ as sketched in Fig. 6.7.

At first, before discussing the results obtained with the proposed boundary element formulation, the analytical 1-d solution of an infinite long column is discussed. Similar to the study in Sect. 7.1, the infinite extension of the column can be approximated by a model of 1000 m length and an observation time short enough that no waves are reflected at the not excited end. But, this will not model geometrical damping as in a half space. In Fig. 7.17, the analytical pressure solution in a distance of 3 m behind the stress excitation point is depicted versus time for Berea sandstone and soil (material data see Table 7.1). Additionally to the results calculated with realistic permeability (solid lines: Berea sandstone $\kappa = 1.9 \cdot 10^{-10}\,\mathrm{m^4/(Ns)}$, soil $\kappa = 3.55 \cdot 10^{-9}\,\mathrm{m^4/(Ns)}$), the pressure solution for an

(a)

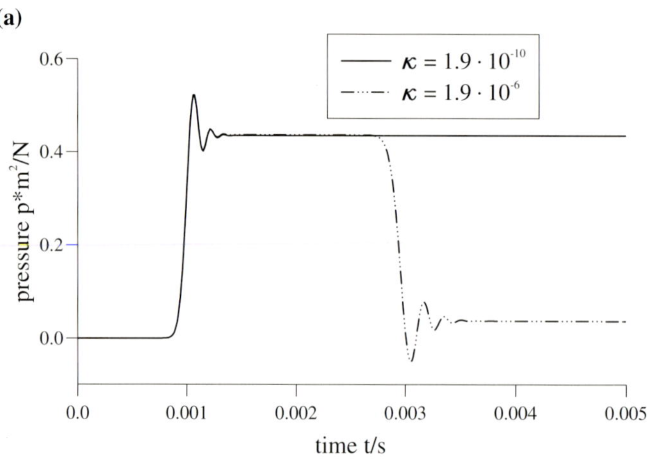

(b)

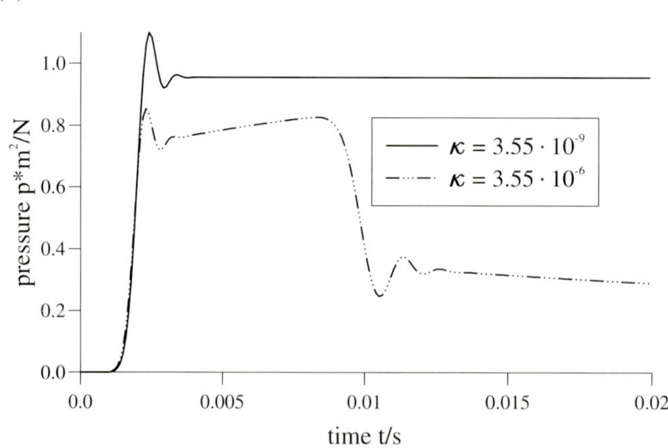

Fig. 7.17. Analytical solution: Pore pressure 3m behind excitation in infinite long column **(a)** Berea sandstone **(b)** soil

increased permeability (dashed lines: Berea sandstone $\kappa = 1.9 \cdot 10^{-6}\,\mathrm{m}^4/(\mathrm{Ns})$, soil $\kappa = 3.55 \cdot 10^{-6}\,\mathrm{m}^4/(\mathrm{Ns})$) is presented. Keeping the intrinsic permeability constant this is equivalent to reducing viscosity, i.e., in the limit an inviscid fluid is assumed. In principle, both materials have the same behavior. A first jump in the pressure solution indicates the arrival of the fast compressional wave. Then, in case of realistic permeability, a constant pressure value is observed. In case of the increased permeabilities, a second jump with negative sign indicates the slow compressional wave. This is in accordance with Biot [31] who has shown that the slow compressional wave has a phase shift of 90 degrees to the fast compressional wave.

Now, the same physical effects will be captured with the 3-d poroelastic boundary element formulation. In Fig. 7.18, the pressure solution of a half space at 3 m below the surface versus time is given for Berea sandstone and for soil. As before, two solutions, one for a realistic value of κ (solid line) and the other for an increased value of κ (dashed line), are presented, respectively. Identical to the analytical solution, two wave fronts are observed arriving at the same time. In case of Berea sandstone, the fast wave arrives at $t \approx 0.001\,\mathrm{s}$ and the slow wave at $t \approx 0.003\,\mathrm{s}$, while in case of soil the arrival times are $t \approx 0.002\,\mathrm{s}$ and $t \approx 0.009\,\mathrm{s}$, respectively. Again, the slow compressional wave is only visible for increased values of κ. In this case, the pressure amplitudes in front of and behind of the slow wave of the 3-d boundary element solution (Fig. 7.18) differ from the amplitudes in the analytical solution (Fig. 7.17). These differences are caused by the geometrical damping. In the 3-d model, the geometrical damping is correctly modeled and, therefore, the amplitudes decrease with increasing distance of the surface. As mentioned above, the 1-d column can not model this behavior leading to same amplitudes at different locations. However, regardless of the model, the ratio between the pressure values in Berea sandstone and in soil is the same. Both calculations give a doubled pressure for soil compared to Berea sandstone.

The most significant difference between 1-d analytical solution and 3-d boundary element solution appears in the case of soil. In the 3-d boundary element solution, a kind of creep behavior is observed for the realistic κ value (see Fig. 7.18b). The pressure solution increases asymptotically to a constant value approximately twice the value as for Berea sandstone. This creep behavior is caused by the dispersion of the compression wave. Whereas for soil this effect is obvious indicating strong damping effects, for Berea sandstone (Fig. 7.18a) nearly no creep is observed indicating very small damping. This is in accordance with larger porosity ϕ of soil compared to Berea sandstone, i.e., more interstitial fluid leads to larger contact areas between solid and fluid and, finally, to an increased dissipated energy. This effect is found in the analytical solution for a slightly increased permeability [164].

When considering the wave front, in both, the analytical and the boundary element solution, "overshooting" is visible. However, this overshooting in Fig. 7.17 is not as strong as in Fig. 7.18. From the analytical solution it is known that this is caused by the time stepping algorithm. Similarly, in the boundary element formulation this large overshooting is caused by numerics and can be influenced con-

(a)

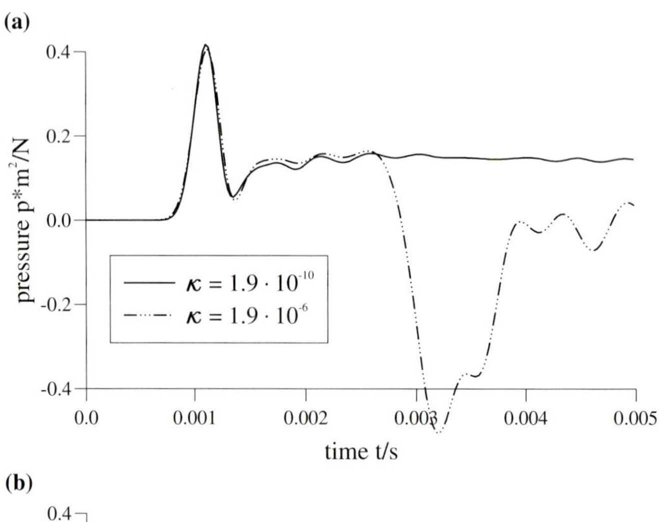

(b)

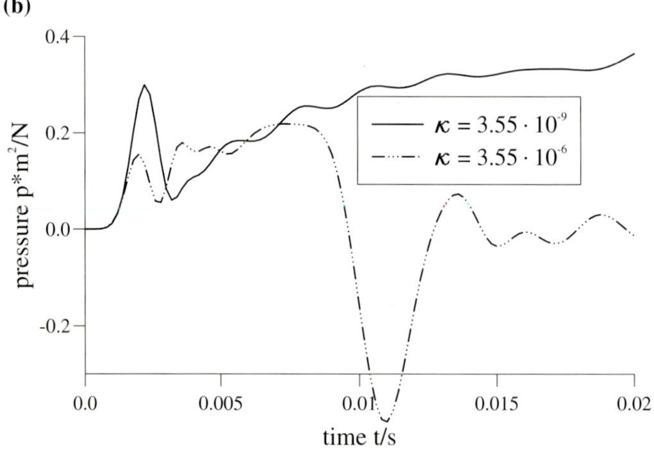

Fig. 7.18. Numerical BE solution: Pore pressure 3 m below surface versus time **(a)** Berea sandstone **(b)** soil

cerning duration and amplitude by the applied multistep method and by the chosen discretization in space and time.

8. Conclusions – Applications

A novel boundary element formulation in time domain has been presented based on the convolution quadrature method. This numerical quadrature formula determines its integration weights from the Laplace transformed fundamental solution and a linear multistep method. Hence, boundary element time-stepping techniques for elastodynamic, viscoelastodynamic, and poroelastodynamic continua have been developed although in case of viscoelasticity and poroelasticity only Laplace domain fundamental solutions are known. So, this method combines the advantage of the Laplace domain with the advantage of a time domain calculation. Finally, wave propagation in a 1-d poroelastic column and in visco- or poroelastic half spaces has been considered.

8.1 Summary

In turn, time-dependent integral equations contain fundamental solutions which are convoluted with time-dependent boundary data. In the presented formulation, this convolution integral is approximated by the convolution quadrature method. The integration weights of this quadrature rule are determined by the Laplace transformed fundamental solution and a linear multistep method. Beside some numerical aspects, this way to establish a time-stepping BE formulation has two main advantages:

1. Only Laplace transformed fundamental solutions are used enabling a time-dependent BE formulation without the knowledge of the time-dependent fundamental solution.
2. The stability of the time-stepping procedure is improved, whereas with different underlying multistep methods different "optimal" time step sizes can be achieved.

Focused on the first advantage, this boundary element method makes it possible to establish time domain boundary element formulations in cases which traditionally are solved in Laplace domain with a subsequent inverse transformation. This traditional procedure is not possible for transient boundary conditions, e.g., contact problems or moving surfaces, and is dependent on a proper choice of some method dependent parameters.

Here, a viscoelastic and a poroelastic BE formulation has been developed without the knowledge of the time-dependent fundamental solutions. Further engineering problem solutions achievable by following the new approach are listed in Table 8.1. There, the focus is on the form of the fundamental solution in which it is

Table 8.1. Availability of fundamental solutions for dynamic problems (* not in closed form, e.g., integrals have to be solved numerically)

fundamental solution	Laplace domain	time domain
Euler-Bernoulli beam	Antes [9]	de Langre [67] *
Timoshenko beam	Antes [8]	Ortner [138]*
Kirchoff plate	Beskos [26]	Benzine, Gamby [22]*
Mindlin plate	Antes [6]	not available
shells	not available	not available
viscoelasticity	Gaul et al. [89]	not available
poroelasticity	Chen [46, 45]	Chen [46, 45]*
thermoelasticity	Nowacki [136]	not available
piezoelectricity	Norris [135]*	Khutoryanski, Sosa[113]*
anisotropic elasticity	Wang, Achenbach [182]*	Wang et al. [183]*
transversely isotropic elasticity	Sáez, Domínguez [154]*	Wang et al. [183]*
elastic half space	Pak [139]*	Triantafyllidis [177]*
layered half space	Apsel, Luco [12]*	Kausel [111]*

provided. For every problem which has at least a fundamental solution in Laplace domain the convolution quadrature based boundary element method leads to a solution. Further, as the complexity of the Laplace domain fundamental solution is mostly less than the corresponding time domain solution, a BE formulation using the convolution quadrature is always advantageous.

The second advantage, listed above, concerning stability was shown in Chap. 4 at the example of an elastodynamic boundary element formulation. As in all BE time-stepping procedures a lower critical time step size exists below which the algorithm becomes unstable. This critical value is approximately ten times smaller as in the "classical" formulation. More important, for a fine enough spatial discretization this value tends to zero, i.e., in the limit no critical lower stability bound exists. Additionally, the underlying multistep method has a strong influence. It was found that a A-stable method with stability in infinity has to be used, e.g., BDF 2.

The convolution quadrature method based BE formulation is introduced and evaluated to model viscoelastic as well as poroelastic continua in the Chaps. 5 and 6. Concerning the spatial and temporal discretization the viscoelastic formulation behaves like the elastodynamic one, whereas the poroelastic formulation needs a much finer spatial discretization as a qualitatively comparable elasto- or viscoelastodynamic calculation.

The influence of the viscoelastic and poroelastic material modeling is studied in Chap. 7 at the example of wave propagation in a half space. Modeling the half space viscoelastic, i.e., taking higher order time derivatives in the stress-strain relation into account, results in a more stiff behavior of the half space compared to an elastic modeling. The poroelastic material model, i.e., dissipation caused by the friction between the elastic solid skeleton and the interstitial viscous pore fluid is introduced, results in the following effects:

- The displacement amplitudes caused by the Rayleigh wave are increased compared to the viscoelastic and the elastic case.
- The Rayleigh wave causes a compressional wave in the pore fluid.
- As expected from theory, the effect of the Rayleigh wave vanishes with depth.
- The second slow compressional wave in the poroelastic medium is found in the limit of an inviscid pore fluid, but for the analyzed materials, Berea sandstone and a soil, for realistic values of the permeability its effect vanishes after a short traveling distance.

These results show that modeling a half space viscoelastic or poroelastic is quite different, i.e., a fluid saturated material should not be modeled viscoelastic.

Summarizing, the proposed boundary element formulation based on the convolution quadrature method combines the advantage of the Laplace domain, i.e., the derivation of a fundamental solution is mostly simpler as in time domain, with the advantage of a time domain calculation, i.e., transient boundary conditions can only be modeled in time domain.

8.2 Outlook on further applications

In the summary given above, several applications of the proposed BE time-stepping technique are given (see Table 8.1). Next, some problems will be sketched which can be solved with the developed poroelastic BE formulation.

Waves in a thermoelastic continuum

The proposed poroelastic formulation can easily be used to solve wave propagation problems in a thermoelastic body. Assuming a linear elastic body with fully coupled thermal effects, the set of governing equations in Laplace domain is given by [136]

$$G\hat{u}_{i,jj} + \left(K + \frac{1}{3}G\right)\hat{u}_{j,ij} - 3K\alpha_t\,\hat{\theta}_{,i} - s^2\rho\hat{u}_i = -\hat{F}_i \tag{8.1}$$

$$\kappa_t\hat{\theta}_{,ii} - sc_\varepsilon\,\hat{\theta} - 3K\alpha_t\theta_0\,s\hat{u}_{i,i} = -\hat{a}_t\ , \tag{8.2}$$

with the solid displacements $\hat{u}_i$ and the temperature $\hat{\theta}$. The other new introduced variables are the coefficient of linear volume expansion α_t, the conductivity κ_t, the specific heat at constant strain c_ε, the average temperature θ_0, and the heat source $\hat{a}_t$. Comparing the above differential equations (8.1) and (8.2) with the set of governing differential equations of a poroelastic continua (6.10) and (6.11), the same structure of the equations is observed. Therefore, identifying the corresponding material parameters by a comparison of the coefficients leads to the thermo-poroelastic analogy [124] (see Table 8.2).

Table 8.2. Analogy between thermoelasticity and poroelasticity in Laplace domain

Variables		Material constants	
thermoelastic	poroelastic	thermoelastic	poroelastic
$\hat{u}_i$	$\hat{u}_i$	$3K\alpha_t$	$\alpha-\beta$
$\hat{\theta}$	$\hat{p}$	κ_t	$\frac{\beta}{s\rho_f}$
$\hat{F}_i$	$\hat{F}_i$	c_ε	$\frac{\phi^2}{R}$
$\hat{a}_t$	$\hat{a}$	ρ	$\rho-\beta\rho_f$

This analogy works only in one direction, from poroelasticity to thermoelasticity but not vice versa [124], because the number of material parameters for both cases in Table 8.2 is not the same. Obviously, for the poroelastic material more parameters are used. Further, due to the dependence of several material "constants" listed in Table 8.2 on the Laplace parameter s, e.g., β, the transformation from one theory to the other is only valid in the Laplace domain.

Implementing this analogy the poroelastic BE time-stepping procedure can be used, e.g., to calculate the time history of the temperature in a body loaded by a mechanical force.

Dam-reservoir systems

The dynamic analysis of dam-reservoir systems subjected to either dynamic forces or seismic waves can be successfully performed by employing the BEM. Especially, the correct treatment of the radiation condition makes the BEM suitable for such problems. For seismic waves a calculation in time domain is necessary.

In Fig. 8.1, the picture of the *Weser*-dam near Eupen in Belgium is given whereas a possible model of this dam-reservoir system is sketched in Fig. 8.2. This cross section is not the real one, it should give only an impression what kind of different materials could appear. The assumed constellation consists of a foundation rock with an upper layer of soil. The retaining wall is made of concrete and the contents of the reservoir is water. Surely, at the bottom of the reservoir a sediment (mud), i.e., a saturated soil with a high porosity, is found. Domínguez has shown that modeling this sediment correctly has an essential influence on the frequency response of the complete system [76].

With the BEM these different layers and also the connection with the concrete dam and the reservoir can be modeled by a substructure technique. However, it is very important to model the mud and the soil as a poroelastic medium in order to take the influence of the pore pressure in these layers into account. Clearly, the water level change the pore pressure and with the pore pressure the material behavior. With

Fig. 8.1. The *Weser*-dam near Eupen in Belgium

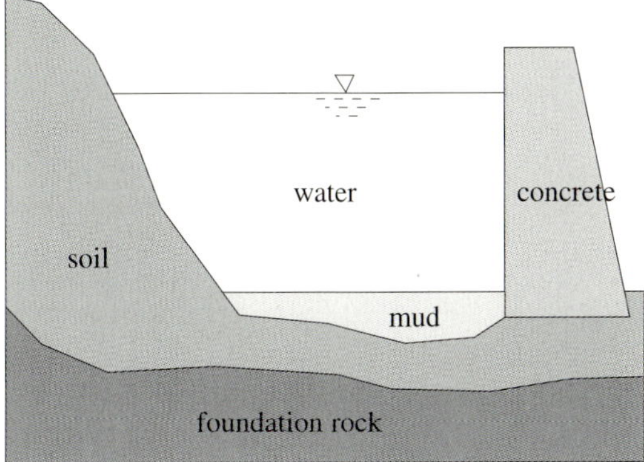

Fig. 8.2. Schematically sketch of a dam-reservoir system cross section

the proposed poroelastic formulation these effects are taken into account, whereas the time-stepping procedure is necessary if the influence of seismic waves or other transient excitations have to be studied.

The curved geometry of the retaining wall makes a 3-d model of the complete structure unavoidable. The symmetries of the real system have to be regarded to find a discretization which can be handled on a computer. In this context, the use of partial geometric symmetries as explained by Bonnet [38] can further reduce the storage and computer time.

Moving loads

Moving loads appear mostly on half spaces caused, e.g., by passing trains. A passing train produces waves in the ground dependent on its speed. The waves or in other words disturbances travel in the half space and influence buildings or people living in the surrounding. Since new high-speed trains are under development an increased interest exists to predict the disturbances caused by such trains in the construction phase of the track.

A rough sketch of a railway-ground system is given in Fig. 8.3. Numerical solu-

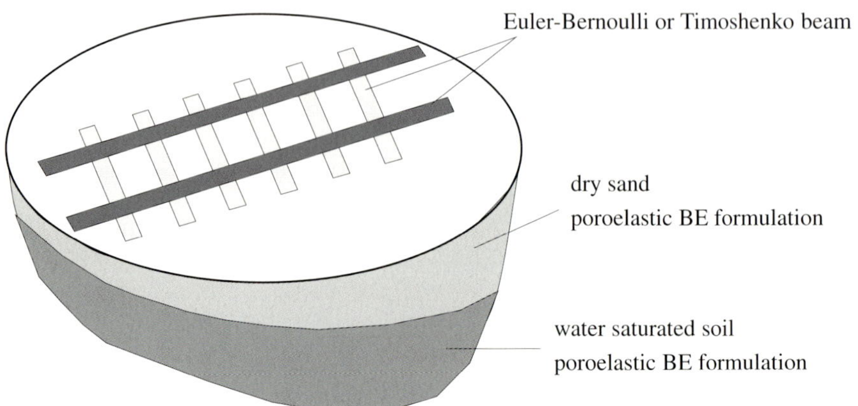

Fig. 8.3. Sketch of a railroad track on a layered soil

tions for the ground vibration related to railway traffic can be found in the literature, e.g., [169, 174], as well as experiments, e.g., [108]. The numerical solutions are often based on modeling the rail as a Timoshenko beam coupled on a half space treated by Green's function [169] or using the FEM [174]. The latter method has problems with the infinite extension of the ground and the other is restricted to cases where Green's function exists, e.g., only horizontal layers can be treated.

Contrary, the proposed BE formulation is not restricted to these cases, e.g., layers with arbitrary shapes and an infinite extension are possible. In order to solve the multi-layered infinite problem with the new approach, the half space can be modeled with the proposed BE formulation using either a viscoelastic material model as in [169] or a poroelastic material model. The different layers are simulated using a substructure technique, where only the interface has to be discretized. If necessary, a foundation rock can be coupled on the water saturated layer by the same technique.

The sleepers are treated by the integral equation for the Euler-Bernoulli beam as described in Chap. 3 or by the modified form of this integral equation for a Timoshenko beam. Converting the integral equation in Chap. 3 to a Timoshenko beam is achieved by simply changing the fundamental solutions [8]. The coupling between the beam and the half space is realized as given by Sheng et al. [169]. As the proposed BE formulation works in time domain, any boundary conditions can be used. Even an uplift of the sleepers can be modeled, taking contact boundary conditions into account.

Bore holes

The last example is especially found in the offshore technique or rather in every hole drilling. Nowadays, tunnels are drilled under streets, buildings, rivers, or other sensitive constructions. In Fig. 8.4[1], a principal sketch of a situation during drilling is depicted with several possible excitations.

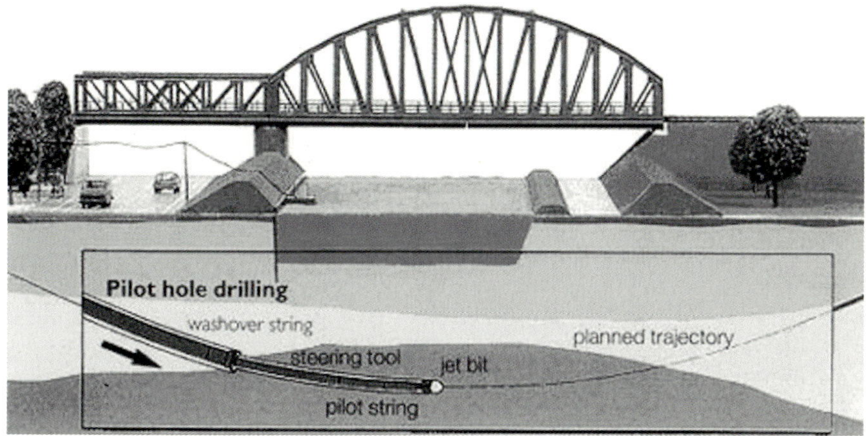

Fig. 8.4. Principal cross section of a half space during a drilling process

There are two points of interest: first, to prevent the bore hole and the technical equipment in the bore hole from damage, and, second, to prevent the surrounding, e.g., buildings, to be damaged by drilling. In both cases, transient excitations have to be considered, e.g., a passing truck on the road above, seismic waves, or an abrupt stop of the drilling due to some change in the material.

The half space material has to be modeled poroelastic because of the drilling fluid used in the bore hole. But, not only the bore hole itself consists of water saturated material most soils are water saturated, especially, in a situation as sketched in Fig. 8.4 where a river flows over the bore hole.

Finally, the arguments in the last two paragraphs enforce to use a time-dependent method which is able to handle poroelastic semi-infinite domains for the prediction of the stresses and displacements in the bore hole and in the surrounding, i.e., the proposed poroelastic BE formulation has to be applied. The layered media of the half space is modeled by the substructure technique. For the bore hole special rotationally symmetric elements have to be developed to take advantage of these symmetries.

[1] This figure is taken from the web-page of the LMR Drilling company `http://www.lmr-drilling.de`

A. Mathematic preliminaries

In the following, some necessary mathematical definitions are recalled. For a rigorous presentation of these definitions the reader is referred to the mathematical literature.

A.1 Distributions or generalized functions

In many engineering fields, physical phenomenon can not be described by functions, e.g., a point force at $x = a$ is everywhere zero except at the point $x = a$. Such a phenomenon is mostly treated with the Dirac "function". However, this is not a function but a distribution or generalized function. Also, sometimes it is necessary to differentiate a piecewise defined function which is only possible in the theory of distribution. This theory was introduced by Schwartz [168]. Here, very briefly the definitions in a non mathematical way are given. Details or more mathematical rigorous treatment can be found, e.g., in [96] or [149]. The following definitions are taken from [149].

First, a more general definition of functions the *linear functional*

$$\langle f, \varphi \rangle = \int_{-\infty}^{\infty} f(x) \varphi(x) \, \mathrm{d}x \tag{A.1}$$

has to be introduced with the *test function* $\varphi(x)$. Contrary to the classical function which associates with every point x a number $y = f(x)$ (the value of f at x), in definition (A.1) the value of the functional $\langle f, \varphi \rangle$ is also a number but represents a "weighted average" of the function f weighted by the test function φ. Such an indirect description of a function is common in engineering. A measuring instrument, such as a voltmeter, does not measure the instantaneous value $f(t_0)$ of the voltage at time t_0, but rather a weighted average over a short time period of time $2T : 1/2T \int_{t_0-T}^{t_0+T} f(t) \varphi(t) \, \mathrm{d}t$, where φ is the characteristic of the measuring instrument.

For our purpose here, the treatment of integral- and differential equations, it will be convenient to restrict the term test function to those functions φ which are continuous, have continuous derivatives of all orders, and vanish outside a certain finite interval, i.e.,

Definition A.1.1 A *test function* φ belongs to the space of C^∞ functions and has a compact support. The *support* of a function $f(x)$ is the closure of the set of points on which $f(x) \neq 0$.

To find a function in C^∞, i.e., the derivatives of all orders are continuous, which vanishes outside a certain finite interval is easy, but for a test function the derivatives of all orders at the boundaries must also be continuous, i.e., they must also vanish. The following function fulfills all conditions and, therefore, could be test a function

$$\varphi(x) = \begin{cases} e^{\frac{1}{x^2-1}} & |x| < 1 \\ 0 & |x| \geq 1 . \end{cases} \tag{A.2}$$

The compact support in equation (A.2) is $[-1,1]$ and all derivatives vanish at $|x| = 1$ [173].

For the following the convergence of a test function is necessary:

Definition A.1.2 A sequence $\varphi_n(x)$ of test functions *converges to zero* ($\varphi_n \to 0$) if:

(a) for each k, the sequence of kth derivatives $\varphi_1^{(k)}, \varphi_2^{(k)}, \ldots$ converges uniformly to zero;

(b) the φ_n have uniformly bounded supports, i.e., there is an interval $[a,b]$, independent of n, such that every $\varphi_n(x)$ vanishes outside of $[a,b]$.

Similarly, it is valid that $\varphi_n \to \varphi$ if the sequence $(\varphi - \varphi_n) \to 0$.

With the definition of the linear functional, now, the derivative of functions f even if they are not continuous at every point can be defined. The derivative of a linear functional and later also of a distribution is given

$$\langle f', \varphi \rangle = \int_{-\infty}^{\infty} f'(x)\,\varphi(x)\,\mathrm{d}x = -\int_{-\infty}^{\infty} f(x)\,\varphi'(x)\,\mathrm{d}x = -\langle f, \varphi' \rangle . \tag{A.3}$$

For a continuously differentiable function f equation (A.3) can be shown with integration by parts. The boundary terms in a partial integration vanish due to the compact support of the test function φ. A very important and often used not continuous function is the *Heaviside* or *Unit step* function

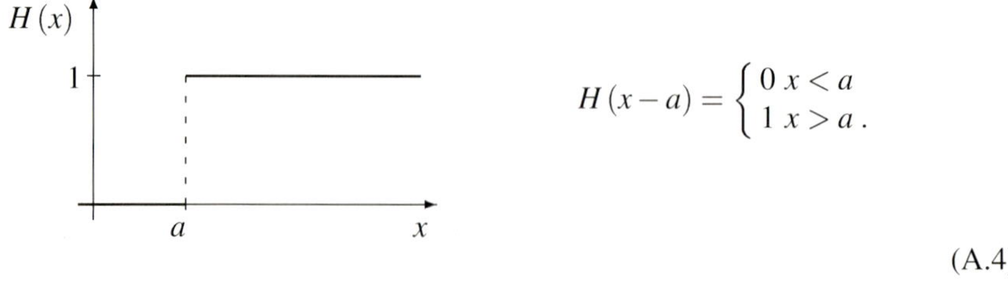

$$H(x-a) = \begin{cases} 0 & x < a \\ 1 & x > a . \end{cases} \tag{A.4}$$

With definition (A.3) a derivative for all x is possible

$$\begin{aligned}\langle H'(x-a),\varphi\rangle &= -\langle H(x-a),\varphi'\rangle = -\int_{-\infty}^{\infty} H(x-a)\,\varphi'(x)\,\mathrm{d}x = -\int_{a}^{\infty} \varphi'(x)\,\mathrm{d}x \\ &= \varphi(a)\ ,\end{aligned} \tag{A.5}$$

using either the property $H(x<a)=0$ and in the last step that a test function vanishes as $x\to\infty$ (compact support). However, the result is not a function in the usual sense, it is a *distribution* or *generalized function*. The distribution in (A.5) is known as the *Dirac* distribution $\delta(x)$ with the known property

$$\int_{-\infty}^{\infty} \delta(x-a)\,\varphi(x)\,\mathrm{d}x = \varphi(a)\ . \tag{A.6}$$

As a consequence of equation (A.5), it is found

$$H'(x-a) = \delta(x-a) \quad \text{and} \quad H(x-a) = \int_{-\infty}^{x-a} \delta(t)\,\mathrm{d}t = \begin{cases} 0 & x<a \\ 1 & x>a\ . \end{cases} \tag{A.7}$$

With (A.6) a distribution was introduced without a definition which will follow now:

Definition A.1.3 A *distribution* T is a mapping from the set of all test functions into the real or complex numbers, such that the following conditions hold:

(a) (Linearity) $\langle T, a\varphi(x)+b\psi(x)\rangle = a\cdot\langle T,\varphi(x)\rangle + b\cdot\langle T,\psi(x)\rangle$ for all test functions φ,ψ and all constants a,b.
(b) (Continuity) If $\varphi_n(x)\to 0$ in the sense defined in definition A.1.2, then $\langle T,\varphi_n(x)\rangle \to 0$

Definition A.1.4 Let $f(x)$ be a piecewise continuous function on the real line. Then we define the distribution T_f corresponding to f by

$$\langle T_f,\varphi(x)\rangle = \int_{-\infty}^{\infty} f(x)\,\varphi(x)\,\mathrm{d}x\ . \tag{A.8}$$

With the last definition the connection between the "normal" functions and the distributions are given. Distributions which are definable in terms of locally integrable functions according to equation (A.8) are called *regular* distributions. All other distributions are called *singular* distributions.

Finally, the properties of the distributions must be presented.

Definition A.1.5 Let S and T be arbitrary distributions. Then we define new distributions $S+T, aT\,(a=\text{constant}), T', T(ax)\,(a\neq 0 \text{ is constant}), T(x-a), g(x)\,T(x)$ (where $g(x)$ is a C^∞ function) by:

$$\langle S+T,\varphi\rangle = \langle S,\varphi\rangle + \langle T,\varphi\rangle \tag{A.9a}$$

$$\langle aT,\varphi\rangle = a\langle T,\varphi\rangle \tag{A.9b}$$

$$\langle T',\varphi\rangle = -\langle T,\varphi'\rangle \tag{A.9c}$$

$$\langle T(ax),\varphi\rangle = |a|^{-1}\langle T,\varphi\left(\frac{x}{a}\right)\rangle \tag{A.9d}$$

$$\langle T(x-a),\varphi\rangle = \langle T,\varphi(x+a)\rangle \tag{A.9e}$$

$$\langle g(x)T(x),\varphi\rangle = \langle T,g(x)\varphi(x)\rangle \tag{A.9f}$$

These, together with convolution (see, e.g., [149]), are the primary operations on distributions. This may seem a rather restrictive list, e.g., there is no definition for the product $S\cdot T$ of two distributions. Unfortunately, the price to be paid for introducing generalized functions (distributions) is that many operations on ordinary functions make no sense in this wider context.

The derivative of a functional was introduced with (A.3) which can be directly applied to distribution as equation (A.9c) shows. For a repeated derivation the rule (A.9c) can be generalized

$$\langle D^k T,\varphi(x)\rangle = (-1)^k \langle T,\varphi^{(k)}(x)\rangle \tag{A.10}$$

with the differential operator D^k denoting the k-th derivative. Thus, equation (A.10) yields the remarkable conclusion that every distribution can be differentiated as often as desired. A distribution can of course be generated by functions which are not differentiable in the ordinary sense, but the theory of distribution provides a way to differentiate such functions in the distributional sense resulting in a distribution, e.g., the Heaviside function.

A final remark must be added. All of the above mentioned can be applied also to n-dimensional distributions.

A.2 Convolution integrals

The convolution integral of two functions $f(t)$ and $g(t)$ is denoted by $*$ and defined as

$$f(t)*g(t) = \int_{-\infty}^{\infty} f(t-\tau)\,g(\tau)\,\mathrm{d}\tau\,. \tag{A.11}$$

For the existence of the convolution integral (A.11) it is sufficient if either f or g is integrable and the other g or f, respectively, bounded. If further both functions vanish for $t<0$ the usual definition is obtained

$$f(t)*g(t) = \begin{cases} \int_0^t f(t-\tau)\,g(\tau)\,\mathrm{d}\tau & t\geq 0 \\ 0 & t<0\,. \end{cases} \tag{A.12}$$

Mostly, the *Riemann* integral definition is used in the convolution integral. However, as for the viscoelastic constitutive equation, a more wider integral definition is necessary. The *Stieltjes convolution* is given

$$f(t) * g(t) = \int_0^t f(t-\tau)\ \mathrm{d}g(\tau)\ , \tag{A.13}$$

where f may have finite jump discontinuities at $t = 0$ while g vanish on $(-\infty, t_0)$ (for some fixed t_0) and be of bounded variation in every closed subinterval of $[t_0, \infty)$ [102].

The above definitions are valid for functions, however, for distributions there is no such a general definition. The reader is referred to the literature, e.g., [149]. Only, the special case of the convolution between the Dirac distribution and any other distribution T with compact support is given

$$\delta * T = T * \delta = T\ . \tag{A.14}$$

Obviously, the convolution of a distribution and the Dirac distribution gives the distribution itself. Because every function can also be represented as a distribution this property can be transferred if T is a "normal" function.

A.3 Laplace transform

A very convenient tool for the solution of differential equations is the Laplace transform. This integral transformation is defined by the *Laplace integral*

$$\hat{f}(s) = \mathscr{L}\{f(t)\} = \int_0^\infty f(t)\,\mathrm{e}^{-st}\mathrm{d}t \tag{A.15}$$

with the complex Laplace variable s for all $f(t<0) = 0$. The Laplace transform of a function f is defined if the Laplace integral (A.15) exists. This condition is fulfilled if the function f is of order $\mathscr{O}\left(\mathrm{e}^{kt}\right)$ for $t \to \infty$, i.e.,

$$|f(t)| \le K\mathrm{e}^{kt}, \quad \text{with} \quad K, k > 0 \ \text{const}, \tag{A.16}$$

is valid and f is piecewise continuous [172]. Then the integral in (A.15) converges for $\Re(s) > k$.

The inverse transformation is achieved by

$$\mathscr{L}^{-1}\left\{\hat{f}(s)\right\} = \frac{1}{2\pi\mathrm{i}} \lim_{R\to\infty} \int_{c-\mathrm{i}R}^{c+\mathrm{i}R} \hat{f}(s)\,\mathrm{e}^{st}\mathrm{d}s = \begin{cases} f(t) & t > 0 \\ 0 & t < 0. \end{cases} \tag{A.17}$$

The integration in (A.17) is performed along a line parallel to the imaginary axis at c. The only condition for c is to be larger than the largest real part of all singularities of $\hat{f}(s)$.

For a plenty of functions the inverse transformation can be found in tables, see, e.g., [71, 172]. If, however, the function is not listed the integral (A.17) has to be solved. This can be done analytically, e.g., using the residue theorem [1] or with the partial fraction decomposition method. Unfortunately, for the practical application, mostly, a numerical method is the only choice. An extensive study of the most numerical transformations can be found in [133] or [53].

A.4 Linear multistep method

An initial value problem is defined by a differential equation of first oder and the initial conditions

$$y'(t) = f(t,y), \quad y(t=0) = y_0 . \tag{A.18}$$

If the inhomogeneity fulfills the *Lipschitz condition* the existence and uniqueness of a solution is given.

After discretization of time t in N time steps Δt a linear multistep method can be applied to the initial value problem (A.18). In contrast to one-step methods, where the numerical solution is obtained solely from the differential equation and the initial value, a k-step method consists of two parts: first, a *starting procedure* which provides the k starting values

$$y(t_n) = y_n \qquad n = 0,1,\dots,k-1 \tag{A.19}$$

(approximations to the exact solution at the points $t_0 + \Delta t, \dots, t_0 + (k-1)\Delta t$) and, second, a multistep (k-step) formula

$$\sum_{j=0}^{k} \alpha_j y_{n+k-j} = \Delta t \sum_{j=0}^{k} \beta_j f\left(t_{n+k-j}, y_{n+k-j}\right) \qquad n = k, k+1, \dots, N \tag{A.20}$$

to obtain an approximation y_n to the exact solution $y(n\Delta t)$. The *characteristic or generating polynomials* are build using the coefficients α_k and β_k

$$\begin{aligned} \rho(\xi) &= \alpha_k \xi^k + \alpha_{k-1}\xi^{k-1} + \dots + \alpha_0 \\ \sigma(\xi) &= \beta_k \xi^k + \beta_{k-1}\xi^{k-1} + \dots + \beta_0 . \end{aligned} \tag{A.21}$$

DEFINITION A.4.1 A multistep method (A.20) is *consistent* if

$$\rho(1) = 0, \qquad \rho'(1) = \sigma(1) . \tag{A.22}$$

is valid.

DEFINITION A.4.2 The linear multistep method (A.20) is called *zero-stable*, if the characteristic polynomial $\rho(\xi)$ satisfies the *root condition*

1. The roots of $\rho(\xi)$ lie on or within the unit circle $|\xi| \leq 1$.
2. The roots on the unit circle $|\xi| = 1$ are simple.

A multistep method is *convergent* if it is zero-stable and consistent [103].

In Chap. 2 the A-stability of multistep methods is used. To define this very restrictive stability criterion, first the stability region has to be defined. Applying the multistep method (A.20) on Dahlquist's test equation $y' = \lambda y$ and using the characteristic polynomials gives

$$\rho(\xi) - z\sigma(\xi) = 0 \quad \text{with} \quad z = \lambda \Delta t\,. \tag{A.23}$$

With this the stability region can be defined [104]:

DEFINITION A.4.3 The set

$$S = \left\{ z \in \mathbb{C}; \quad \begin{array}{l} \text{all roots } \xi_j(z) \text{ of (A.23) satisfy } |\xi_j(z)| \leq 1, \text{ multiple} \\ \text{roots satisfy } |\xi_j(z)| < 1 \end{array} \right\}$$

is called the *stability region* or *stability domain* or *region of absolute stability* of method (A.20).

Now, the A- and the $A(\alpha)$-stability can be defined:

DEFINITION A.4.4 The multistep method (A.20) is called *A-stable* if

$$\mathbb{C}^- \subset S$$

and it is called *$A(\alpha)$-stable* with $\alpha \in \left[0, \frac{\pi}{2}\right]$ if

$$\left\{ z \in \mathbb{C}^- \text{ with } |\arg(z) - \pi| \leq \alpha \right\} \subset S\,.$$

Additional to the A-stability in Chap. 2 "stability in infinity" is required. For one-step methods this stability is called L-stability. A similar definition is not found in the literature for multistep methods. However, this property is important as well for multistep methods. The *stability in infinity* is equal to the requirement that in (A.23) for $|z| \to \infty$ all k zeros of the characteristic equation tend to zero.

B. BEM details

B.1 Fundamental solutions

For the BEM it is essential to have a fundamental solution. These solutions are defined for an arbitrary differential operator L as follows [173]:

DEFINITION B.1.1 A *fundamental solution* $E(\mathbf{x},\mathbf{y})$ for L with pole at $\mathbf{y}$ is a solution of the equation

$$Lu = \delta(\mathbf{x}-\mathbf{y}) \ .$$

If L has constant coefficients the fundamental solution has the property $E(\mathbf{x},\mathbf{y}) = E(\mathbf{x}-\mathbf{y},\mathbf{0})$

From the definition B.1.1 it is clear that this equation and, therefore, the fundamental solution has to be interpreted in the sense of a distribution.

Physically interpreted the fundamental solution is the response due to an impulse loading (point load) at the point $\mathbf{y}$ of a system governed by the differential operator L.

B.1.1 Visco- and elastodynamic fundamental solutions

For completeness, here, the visco- and elastodynamic fundamental solutions in Laplace domain are recalled (see [61]). Due to the usage of the convolution quadrature method no time-dependent fundamental solutions are necessary. The elastodynamic displacement fundamental solution is

$$\begin{aligned}\hat{U}_{ij}(\mathbf{x},\mathbf{y},s) = \frac{1}{4\pi\rho}\Bigg\{ & \frac{3r_{,i}r_{,j}-\delta_{ij}}{r^3}\left[\frac{s\frac{r}{c_1}+1}{s^2}\mathrm{e}^{-\frac{r}{c_1}s} - \frac{s\frac{r}{c_2}+1}{s^2}\mathrm{e}^{-\frac{r}{c_2}s}\right] \\ & + \frac{r_{,i}r_{,j}}{r}\left[\frac{1}{c_1^2}\mathrm{e}^{-\frac{r}{c_1}s} - \frac{1}{c_2^2}\mathrm{e}^{-\frac{r}{c_2}s}\right] + \frac{\delta_{ij}}{rc_2^2}\mathrm{e}^{-\frac{r}{c_2}s}\Bigg\}\end{aligned} \tag{B.1}$$

and the corresponding traction solution

$$
\begin{aligned}
\hat{T}_{ij}(\mathbf{x},\mathbf{y},s) &= \rho\left(c_1^2-2c_2^2\right)\hat{U}_{kj,k}n_i+\rho c_2^2\left(\hat{U}_{ij,k}+\hat{U}_{kj,i}\right)n_k \\
&= \frac{1}{4\pi}\left\{\frac{6c_2^2}{r^2}\left(r_{,i}n_j+r_{,j}n_i+r_{,n}\left(\delta_{ij}-5r_{,i}r_{,j}\right)\right)\right. \\
&\qquad \left[\frac{e^{-\frac{r}{c_1}s}}{c_1^2}\left(\frac{c_1}{rs}+\frac{c_1^2}{r^2s^2}\right)-\frac{e^{-\frac{r}{c_2}s}}{c_2^2}\left(\frac{c_2}{rs}+\frac{c_2^2}{r^2s^2}\right)\right] \\
&\quad +\frac{e^{-\frac{r}{c_1}s}}{r^2c_1^2}\left[2c_2^2\left(2r_{,j}n_i+r_{,i}n_j-6r_{,i}r_{,j}r_{,n}+\delta_{ij}r_{,n}\right)-c_1^2r_{,j}n_i\right] \\
&\quad +\frac{e^{-\frac{r}{c_2}s}}{r^2}\left[12r_{,i}r_{,j}r_{,n}-2r_{,j}n_i-3r_{,i}n_j-3\delta_{ij}r_{,n}\right] \\
&\quad -\frac{e^{-\frac{r}{c_1}s}s}{rc_1^3}\left[c_1^2r_{,j}n_i+2c_2^2\left(r_{,i}r_{,j}r_{,n}-r_{,j}n_i\right)\right] \\
&\quad \left.+\frac{e^{-\frac{r}{c_2}s}s}{rc_2}\left[2r_{,i}r_{,j}r_{,n}-\delta_{ij}r_{,n}-r_{,i}n_j\right]\right\}
\end{aligned}
\tag{B.2}
$$

with $r_i = x_i - y_i$ and $r = \sqrt{r_i r_i}$.

With the elastic-viscoelastic correspondence principle the elastodynamic fundamental solutions (B.1) and (B.2) are converted to the viscoelastic ones. This means that the elastic material constants K and G have to be exchanged by the correspondence (5.17). These material constants appear in the elastodynamic fundamental solutions only in the elastic wave velocities

$$
c_1^2 = \frac{K+\frac{4}{3}G}{\rho} \qquad c_2^2 = \frac{G}{\rho}\,. \tag{B.3}
$$

Inserting there the elastic-viscoelastic correspondence (5.17) results in the expressions

$$
c_{1v}^2(s) = \frac{1}{\rho}\left(K\frac{1+q^H s^{\alpha^H}}{1+p^H s^{\alpha^H}}+\frac{4}{3}G\frac{1+q^D s^{\alpha^D}}{1+p^D s^{\alpha^D}}\right) \qquad c_{2v}^2(s) = \frac{G}{\rho}\frac{1+q^D s^{\alpha^D}}{1+p^D s^{\alpha^D}}\,. \tag{B.4}
$$

These expressions should not be mismatched with viscoelastic wave velocities. The viscoelastic wave velocities are mostly defined with the initial moduli [55]

$$
c_1^2 = \frac{1}{\rho}\left(K\frac{q^H}{p^H}+\frac{4}{3}G\frac{q^D}{p^D}\right) \qquad c_2^2 = \frac{G}{\rho}\frac{q^D}{p^D}\,. \tag{B.5}
$$

Exchanging in (B.1) and (B.2) the elastic wave velocities c_1 and c_2 with the expressions c_{1v} and c_{2v} of equation (B.4), respectively, yields the viscoelastic fundamental solutions.

B.1.2 Poroelastodynamic fundamental solutions

The explicit expressions of the poroelastodynamic fundamental solutions are given in the following. The four elements of the matrix $\mathbf{G}$ are the displacements caused by a Dirac force in the solid:

$$\hat{U}^s_{ij} = \frac{1}{4\pi r\left(\rho - \beta\rho_f\right)s^2}\left[R_1 \frac{\lambda_4^2 - \lambda_2^2}{\lambda_1^2 - \lambda_2^2} \mathrm{e}^{-\lambda_1 r} - R_2 \frac{\lambda_4^2 - \lambda_1^2}{\lambda_1^2 - \lambda_2^2} \mathrm{e}^{-\lambda_2 r} + \left(\delta_{ij}\lambda_3^2 - R_3\right)\mathrm{e}^{-\lambda_3 r}\right] \tag{B.6}$$

with $R_k = \left(3r_{,i}r_{,j} - \delta_{ij}\right)/r^2 + \lambda_k\left(3r_{,i}r_{,j} - \delta_{ij}\right)/r + \lambda_k^2 r_{,i}r_{,j}$ and $\lambda_4^2 = s^2\left(\rho - \beta\rho_f\right)/\left(K + 4/3G\right)$. The pressure caused by the same load is

$$\hat{P}^s_j = \frac{(\alpha - \beta)\, s\rho_f r_{,j}}{4\pi\beta\left(K + \frac{4}{3}G\right) r\left(\lambda_1^2 - \lambda_2^2\right)}\left[\left(\lambda_1 + \frac{1}{r}\right)\mathrm{e}^{-\lambda_1 r} - \left(\lambda_2 + \frac{1}{r}\right)\mathrm{e}^{-\lambda_2 r}\right] . \tag{B.7}$$

For a Dirac source in the fluid the respective displacement solution is

$$\hat{U}^f_i = s\hat{P}^s_i \tag{B.8}$$

and the pressure

$$\hat{P}^f = \frac{s\rho_f}{4\pi r\beta\left(\lambda_1^2 - \lambda_2^2\right)}\left[\left(\lambda_1^2 - \lambda_4^2\right)\mathrm{e}^{-\lambda_1 r} - \left(\lambda_2^2 - \lambda_4^2\right)\mathrm{e}^{-\lambda_2 r}\right] . \tag{B.9}$$

The roots λ_i, $i = 1,2,3$ of the characteristic equation (6.27) are recalled here:

$$\lambda_{1,2}^2 = \frac{1}{2}\left[\frac{\phi^2 s^2 \rho_f}{\beta R} + \frac{s^2\left(\rho - \beta\rho_f\right)}{K + \frac{4}{3}G} + \frac{s^2\rho_f\left(\alpha - \beta\right)^2}{\beta\left(K + \frac{4}{3}G\right)} \pm \sqrt{\left(\frac{\phi^2 s^2 \rho_f}{\beta R} + \frac{s^2\left(\rho - \beta\rho_f\right)}{K + \frac{4}{3}G} + \frac{s^2\rho_f\left(\alpha - \beta\right)^2}{\beta\left(K + \frac{4}{3}G\right)}\right)^2 - 4\frac{s^4\phi^2\rho_f\left(\rho - \beta\rho_f\right)}{\beta R\left(K + \frac{4}{3}G\right)}}\right]$$

$$\lambda_3^2 = \frac{s^2\left(\rho - \beta\rho_f\right)}{G} .$$

In the derivation of the poroelastodynamic boundary integral equation (6.39) several abbreviations (6.40) corresponding to an "adjoint" traction or flux are introduced. First, the "adjoint" traction solution is presented. However, due to the extensive expression only parts are given

$$\hat{T}^s_{ij} = \left[\left(\left(K - \frac{2}{3}G\right)\hat{U}^s_{kj,k} + \alpha s\hat{P}^s_j\right)\delta_{i\ell} + G\left(\hat{U}^s_{ij,\ell} + \hat{U}^s_{\ell j,i}\right)\right]n_\ell \tag{B.10}$$

$$\hat{U}^s_{kj,k}\delta_{i\ell}n_\ell = \frac{r_{,j}n_i}{4\pi r s^2\left(\rho - \beta\rho_f\right)\left(\lambda_1^2 - \lambda_2^2\right)}\left[\mathrm{e}^{-\lambda_1 r}\left(\frac{1}{r} + \lambda_1\right)\lambda_1^2\left(\lambda_2^2 - \lambda_4^2\right) - \mathrm{e}^{-\lambda_2 r}\left(\frac{1}{r} + \lambda_2\right)\lambda_2^2\left(\lambda_1^2 - \lambda_4^2\right)\right]$$

$$
\begin{aligned}
&(\hat{U}^s_{ij,\ell}+\hat{U}^s_{\ell j,i})n_\ell \\
&=\frac{1}{4\pi r s^2(\rho-\beta\rho_f)}\left[\frac{R_56}{r^3}\left(\frac{\lambda_4^2-\lambda_2^2}{\lambda_1^2-\lambda_2^2}\mathrm{e}^{-\lambda_1 r}-\frac{\lambda_4^2-\lambda_1^2}{\lambda_1^2-\lambda_2^2}\mathrm{e}^{-\lambda_2 r}-\mathrm{e}^{-\lambda_3 r}\right)\right.\\
&\quad+\frac{R_56}{r^2}\left(\frac{\lambda_4^2-\lambda_2^2}{\lambda_1^2-\lambda_2^2}\lambda_1\mathrm{e}^{-\lambda_1 r}-\frac{\lambda_4^2-\lambda_1^2}{\lambda_1^2-\lambda_2^2}\lambda_2\mathrm{e}^{-\lambda_2 r}-\lambda_3\mathrm{e}^{-\lambda_3 r}\right)\\
&\quad+\frac{R_62}{r}\left(\frac{\lambda_4^2-\lambda_2^2}{\lambda_1^2-\lambda_2^2}\lambda_1^2\mathrm{e}^{-\lambda_1 r}-\frac{\lambda_4^2-\lambda_1^2}{\lambda_1^2-\lambda_2^2}\lambda_2^2\mathrm{e}^{-\lambda_2 r}-\lambda_3^2\mathrm{e}^{-\lambda_3 r}\right)\\
&\quad-2r_{,n}r_{,i}r_{,j}\left(\frac{\lambda_4^2-\lambda_2^2}{\lambda_1^2-\lambda_2^2}\lambda_1^3\mathrm{e}^{-\lambda_1 r}-\frac{\lambda_4^2-\lambda_1^2}{\lambda_1^2-\lambda_2^2}\lambda_2^3\mathrm{e}^{-\lambda_2 r}-\lambda_3^3\mathrm{e}^{-\lambda_3 r}\right)\\
&\quad\left.-\lambda_3^2(\delta_{ij}r_{,n}+r_{,i}n_j)\left(\lambda_3+\frac{1}{r}\right)\mathrm{e}^{-\lambda_3 r}\right]
\end{aligned}
$$

with $R_5 = r_{,j}n_i + r_{,i}n_j + r_{,n}(\delta_{ij} - 5r_{,i}r_{,j})$ and $R_6 = r_{,j}n_i + r_{,i}n_j + r_{,n}(\delta_{ij} - 6r_{,i}r_{,j})$. The other explicit expressions are:

$$
\begin{aligned}
\hat{Q}^s_j = \frac{n_i}{4\pi r(\rho-\beta\rho_f)s^2}&\left[\frac{\mathrm{e}^{-\lambda_1 r}}{\lambda_1^2-\lambda_2^2}R_1\left(\beta\lambda_2^2-\alpha\lambda_4^2\right)-\frac{\mathrm{e}^{-\lambda_2 r}}{\lambda_1^2-\lambda_2^2}R_2\left(\beta\lambda_1^2-\alpha\lambda_4^2\right)\right.\\
&\left.+\beta\mathrm{e}^{-\lambda_3 r}\left(R_3-\delta_{ij}\lambda_3^2\right)\right]
\end{aligned}
\tag{B.11}
$$

$$
\begin{aligned}
\hat{T}^f_i =&\frac{s^2\rho_f}{4\pi r\beta\left(\lambda_1^2-\lambda_2^2\right)}\left[\frac{n_j(\alpha-\beta)2G}{K+\frac{4}{3}G}\left(R_2\mathrm{e}^{-\lambda_2 r}-R_1\mathrm{e}^{-\lambda_1 r}\right)\right.\\
&+n_i\mathrm{e}^{-\lambda_2 r}\left(\frac{(\alpha-\beta)\left(K-\frac{2}{3}G\right)}{K+\frac{4}{3}G}\left(\frac{2}{r^2}+\frac{2\lambda_2}{r}+\lambda_2^2\right)-\alpha\left(\lambda_2^2-\lambda_4^2\right)\right)\\
&\left.-n_i\mathrm{e}^{-\lambda_1 r}\left(\frac{(\alpha-\beta)\left(K-\frac{2}{3}G\right)}{K+\frac{4}{3}G}\left(\frac{2}{r^2}+\frac{2\lambda_1}{r}+\lambda_1^2\right)-\alpha\left(\lambda_1^2-\lambda_4^2\right)\right)\right]
\end{aligned}
\tag{B.12}
$$

$$
\begin{aligned}
\hat{Q}^f = \frac{r_{,n}}{4\pi r\left(\lambda_1^2-\lambda_2^2\right)}&\left[\left(\lambda_2+\frac{1}{r}\right)\left(\lambda_2^2-\lambda_4^2\frac{\rho-\alpha\rho_f}{\rho-\beta\rho_f}\right)\mathrm{e}^{-\lambda_2 r}\right.\\
&\left.-\left(\lambda_1+\frac{1}{r}\right)\left(\lambda_1^2-\lambda_4^2\frac{\rho-\alpha\rho_f}{\rho-\beta\rho_f}\right)\mathrm{e}^{-\lambda_1 r}\right]
\end{aligned}
\tag{B.13}
$$

B.2 "Classical" time domain BE formulation

The time stepping procedure proposed by Mansur [126] and later extended to non-zero initial conditions by Antes [4] approximates the spatial behavior with polynomial shape functions. Additionally, the time t is approximated also by polynomials, mostly, linear for displacement and constant for traction after discretizing t in N equal time steps Δt. There is one different approach from Karabalis and Rizos [109] where spline functions are used to approximate the time history.

Using the classical procedure with linear and constant polynomials for the displacement and traction, respectively, and with the nodal values U_{if}^{em} and T_{if}^{em} at the time and spatial collocation points the ansatz functions are

$$\begin{aligned} u_i(\tau,\mathbf{x}) &= \sum_{e=1}^{E}\sum_{f=1}^{F}\sum_{m=1}^{N} N_e^f(\mathbf{x})\left(U_{if}^{em-1}\frac{t_m-\tau}{\Delta t}+U_{if}^{em}\frac{\tau-t_{m-1}}{\Delta t}\right) \\ t_i(\tau,\mathbf{x}) &= \sum_{e=1}^{E}\sum_{f=1}^{F}\sum_{m=1}^{N} N_e^f(\mathbf{x})\,T_{if}^{em} \end{aligned} \tag{B.14}$$

Inserting these ansatz functions in the integral equation (4.17) yields

$$\begin{aligned} c_{ij}(\mathbf{y})u_i(\mathbf{y},t) = \sum_{e=1}^{E}\sum_{f=1}^{F}\sum_{m=1}^{N}\Bigg\{&\int_\Gamma N_e^f(\mathbf{x})\int_{t_{m-1}}^{t_m} U_{ij}(\mathbf{x},\mathbf{y},t-\tau)\,\mathrm{d}\tau\,\mathrm{d}\Gamma \\ &-\oint_\Gamma N_e^f(\mathbf{x})\int_{t_{m-1}}^{t_m} T_{ij}(\mathbf{x},\mathbf{y},t-\tau)\left(U_{if}^{em-1}\frac{t_m-\tau}{\Delta t}+U_{if}^{em}\frac{\tau-t_{m-1}}{\Delta t}\right)\mathrm{d}\tau\mathrm{d}\Gamma\Bigg\} \\ &+\int_\Omega b_i(t,\mathbf{x}) * U_{ij}(t,\mathbf{x},\mathbf{y})\,\mathrm{d}\Omega\,. \end{aligned} \tag{B.15}$$

Due to the properties of the time-dependent fundamental solutions – they consist of Dirac distributions and Heaviside functions – the time integration within each time step is performed analytically. For the spatial integration Gaussian quadrature formulas are used except when $\mathbf{x}$ approaches $\mathbf{y}$. In this singular case, the same regularization methods as described in Chap. 4.2 are applied. Point collocation and the properties of the fundamental solutions – causality and independence on t but dependence on the difference $t-\tau$ – yield a time stepping procedure

$$\mathbf{C}^1\mathbf{d}^n = \mathbf{D}^1\bar{\mathbf{d}}^n + \sum_{k=2}^{\bar{n}}\left(\mathbf{U}^k\mathbf{t}^{n-k+1}-\mathbf{T}^k\mathbf{u}^{n-k+1}\right)+\mathbf{b}^n \quad n=1,2,\ldots,N \tag{B.16}$$

with the matrix of the time integrated fundamental solutions of the tractions $\mathbf{T}^k$ and displacements $\mathbf{U}^k$. In the matrices $\mathbf{C}^1$ and $\mathbf{D}^1$ are the time integrated fundamental solutions of the first time step assembled concerning the unknown boundary data $\mathbf{d}$ and the known boundary data $\bar{\mathbf{d}}$, respectively. The vector $\mathbf{b}^n$ denotes the integrated volume forces. The upper limit of the sum $\bar{n}$ is determined by the fact that in 3-d the fundamental solutions are zero after passing of the shear wave, i.e., $\bar{n} = \min(n, r_{max}/(c_2\Delta t)+2)$ in 3-d and $\bar{n}=n$ in 2-d.

For more details about this direct approach in time domain see e.g., [76] or [127].

Notation Index

$\mathbf{A}$	matrix or matrix differential operator
$\mathbf{A}^*$	adjoint operator matrix
$\mathbf{A}^{co}$	matrix of cofactors
$\mathbf{a}$	vector
$a(t)$	source in the pore fluid
α	Biot's effective stress coefficient, parameter $\Re(s)$ in Durbin's inverse transformation
α^D, α^H	fractional exponent in viscoelastic constitutive equation
C_{ijkl}	elasticity tensor
c_{ij}	integral free term
$\gamma(z) = \frac{\rho(z)}{\sigma(z)}$	characteristic function of the multistep method
Δt	time step size
$\delta(t)$	Dirac distribution
E	Young's modulus
EI	flexural rigidity
e_{ij}	component of the deviatoric part of the strain tensor
ε_{ij}	component of the strain tensor, poroelastic: solid strain
ε	error bound for calculating $\hat{f}(s)$
F_i	component of bulk body force
G	shear modulus
$H(t)$	Heaviside- or Unit Step function
$\mathbf{I}$	identity matrix
$\mathrm{i} = \sqrt{-1}$	imaginary number
K	compression modulus
K_s	compression modulus of the solid grains
K_f	compression modulus of the fluid
κ	permeability
ℓ	length of the beam or column
L	amount of integration steps for determining ω_n
M	bending moment
N	total amount of time steps

$N_e^f(\mathbf{x})$	spatial shape function
ν	Poisson's ratio
p	pore pressure
p^D, p^H, q^D, q^H	viscoelastic material constants
ϕ	porosity
Q	shear force
q	specific flux
ρ	density, poroelastic: bulk density
ρ_f	fluid density
ρ_s	solid density
$r = \|\mathbf{x}-\mathbf{y}\|$	distance of the points $\mathbf{x}$ and $\mathbf{y}$
$\mathscr{R}$	radius of a circle in the domain of analyticity of $\hat{f}\left(\frac{\gamma(z)}{\Delta t}\right)$
$s \in \mathbb{C}$	complex Laplace variable
σ_{ij}	component of the stress tensor, poroelastic: total stress tensor
s_{ij}	component of the deviatoric part of the stress tensor
t, τ	time
t_i	traction vector component
T_{ij}	fundamental solutions of the traction
u_i	displacement vector component
U_{ij}	fundamental solutions of the displacement
v_i	relative fluid to solid displacement component
w	beam deflection
ω_n	integration weight
Ω, Γ	domain with boundary
ζ	variation of fluid volume per unit reference volume
$\\|\cdot\\|$	norm of $\cdot$
$()^*$	fundamental solution for the beam
$()^{(i)}$	i-th partial derivative
$\frac{\partial}{\partial()}$	partial derivative
$()_{,i}$	partial derivative with respect to x_i
$\dot{()}$	time derivative
$f(t) * g(t)$	convolution of the functions $f(t)$ and $g(t)$
δ_{ij}	Kronecker symbol
$f(x) = \mathscr{O}(g(x))$	Landau symbol: $\lim_{x\to x_0}\left\|\frac{f(x)}{g(x)}\right\| < C$
$\mathscr{L}\{f(t)\}, \hat{f}(s)$	Laplace transform of f
$\mathscr{L}^{-1}\{\hat{f}(s)\}$	inverse Laplace transform of f
$\Re(s), \Im(s)$	real and imaginary part of complex number s

References

1. Ablowitz, M.J.; Fokas, A.S.: *Complex Variables: Introduction and Applications*. Cambridge Texts in Applied Mathematics. Cambridge University Press, Cambridge, 1997.
2. Achenbach, J.D.: *Wave Propagation in Elastic Solids*. North Holland, 1980.
3. Ahmad, S.; Manolis, G.D.: Dynamic Analysis of 3-D Structures by a Transformed Boundary Element Method. *Computational Mechanics*, **2**, 185–196, 1987.
4. Antes, H.: A Boundary Element Procedure for Transient Wave Propagations in Two-dimensional Isotropic Elastic Media. *Finite Elements in Analysis and Design*, **1**, 313–322, 1985.
5. Antes, H.: *Anwendungen der Methode der Randelemente in der Elastodynamik und der Fluiddynamik*. Mathematische Methoden in der Technik 9. B. G. Teubner, Stuttgart, 1988.
6. Antes, H.: Static and Dynamic Analysis of Reissner-Mindlin Plates. In *Boundary Element Analysis of Plates and Shells*. (Beskos, D.E., Ed.), Springer Series in Computational Mechanics, Springer-Verlag, Berlin, Heidelberg, New York, 312–340, 1991.
7. Antes, H.: A Boundary Element Formulation for the Dead Weight Integral in 2-D Elastodynamic BEM. *Mechanics Research Communications*, **19**(4), 273–278, 1992.
8. Antes, H.: Fundamental Solution and Integral Equations for Timoshenko Beams. *Computers & Structures*, (submitted).
9. Antes, H.: On the Completeness of Fundamental Solutions for Harmonic Problems. *Engineering Analysis with Boundary Elements*, (submitted).
10. Antes, H.; Jäger, M.: On Stability and Efficiency of 3D Acoustic BE Procedures for Moving Noise Sources. In *Computational Mechanics, Theory and Applications*. (Atluri, S.N.; Yagawa, G.; Cruse, T.A., Eds.), Vol. 2, Heidelberg, Springer-Verlag, 3056–3061, 1995.
11. Antes, H.; Panagiotopoulos, P.D.: *The Boundary Integral Approach to Static and Dynamic Contact Problems — Equality and Inequality Methods*. Int. Series of Numerical Mathematics 108. Birkhäuser, Basel, 1992.
12. Apsel, R.J.; Luco, J.E.: On the Green's Functions for a Layered Half-Space. Part II. *Bulletin of the Seismological Society of America*, **73**(4), 931–951, 1983.
13. Ayres, F.: *Theory and Problems of Matrices*. Schaum's Outline Series. McGraw Hill Book Company, 1962.
14. Badiey, M.; Cheng, A.H.-D.; Mu, Y.: From Geology to Geoacoustics – Evaluation of Biot-Stoll Sound Speed and Attenuation for Shallow Water Acoustics. *Journal of the Acoustical Society of America*, **103**(1), 309–320, 1998.
15. Badmus, T.; Cheng, A.H.-D.; Grilli, S.: A Laplace-Transform Based Three-Dimensional BEM for Poroelasticity. *International Journal for Numerical Methods in Engineering*, **36**(1), 67–85, 1993.
16. Bagley, R.L.; Torvik, P.J.: A Theoretical Basis for the Application of Fractional Calculus to Viscoelasticity. *Journal of Rheology*, **27**(3), 201–210, 1983.
17. Bagley, R.L.; Torvik, P.J.: On the Fractional Calculus Model of Viscoelastic Behaviour. *Journal of Rheology*, **30**(1), 133–155, 1986.

18. Banerjee, P.K.; Ahmad, S.; Wang, H.C.: Advanced Development of BEM for Elastic and Inelastic Dynamic Analysis of Solids. In *Industrial Applications of Boundary Element Methods.* (Banerjee, P.K.; Wilson, R.B., Eds.), Developments in Boundary Element Methods, Elsevier, London, 77–117, 1989.
19. Banerjee, P.K.; Butterfield, R.: Boundary Element Methods in Geomechanics. In *Finite Elements in Geomechanics.* (Gudehus, G., Ed.), London, J. Wiley and Sons, 529–570, 1977.
20. Banerjee, P.K.; Butterfield, R.: *Boundary Element Methods in Engineering Science.* McGraw Hill, London, 1981.
21. Bathe, K.-J.: *Finite Element Procedures in Engineering Analysis.* Prentice Hall, Englewood Cliffs, NJ, 1982.
22. Benzine, G.; Gamby, D.: Étude des Mouvements Transitoires de Flexion d'une Plaque par la Méthode des Équations Intégrales de Frontière. *Journal de Mécanique Appliquée*, **1**(3), 451–466, 1982.
23. Beskos, D. E.: Boundary Element Methods in Dynamic Analysis: Part II (1986-1996). *Applied Mechanics Review*, **50**(3), 149–197, 1997.
24. Beskos, D.E.: Boundary Element Methods in Dynamic Analysis. *Applied Mechanics Review*, **40**(1), 1–23, 1987.
25. Beskos, D.E.: Introduction to Boundary Element Methods. In *Boundary Element Methods in Mechanics.* (Beskos, D.E., Ed.), Mechanics and Mathematical Methods, North-Holland, Amsterdam, New York, 1–21, 1987.
26. Beskos, D.E.: Dynamic Analysis of Plates. In *Boundary Element Analysis of Plates and Shells.* (Beskos, D.E., Ed.), Springer Series in Computational Mechanics, Springer-Verlag, Berlin, Heidelberg, New York, 35–92, 1991.
27. Bettess, P.: *Infinite Elements.* Penshaw Press, 1992.
28. Biot, M.A.: General Theory of Three-Dimensional Consolidation. *Journal of Applied Physics*, **12**, 155–164, 1941.
29. Biot, M.A.: Theory of Elasticity and Consolidation for a Porous Anisotropic Solid. *Journal of Applied Physics*, **26**, 182–185, 1955.
30. Biot, M.A.: Theory of Deformation of a Porous Viscoelastic Anisotropic Solid. *Journal of Applied Physics*, **27**(5), 459–467, 1956.
31. Biot, M.A.: Theory of Propagation of Elastic Waves in a Fluid-Saturated Porous Solid.I. Low-Frequency Range. *Journal of the Acoustical Society of America*, **28**(2), 168–178, 1956.
32. Biot, M.A.: Theory of Propagation of Elastic Waves in a Fluid-Saturated Porous Solid.II. Higher Frequency Range. *Journal of the Acoustical Society of America*, **28**(2), 179–191, 1956.
33. Birgisson, B.; Siebrits, E.; Peirce, A.P.: Elastodynamic Direct Boundary Element Methods with Enhanced Numerical Stability Properties. *International Journal for Numerical Methods in Engineering*, **46**, 871–888, 1999.
34. Boltzmann, L.: Zur Theorie der elastischen Nachwirkungen. *Sitzungsbericht der Akademie der Wissenschaften (Wien): Mathematisch–Naturwissenschaftlichen Klasse*, **70 (2)**, 275–300, 1874.
35. Bonnet, G.: Basic Singular Solutions for a Poroelastic Medium in the Dynamic Range. *Journal of the Acoustical Society of America*, **82**(5), 1758–1762, 1987.
36. Bonnet, G.; Auriault, J.-L.: Dynamics of Saturated and Deformable Porous Media: Homogenization Theory and Determination of the Solid-Liquid Coupling Coefficients. In *Physics of Finely Divided Matter.* (Boccara, N.; Daoud, M., Eds.), Springer Verlag, Berlin, 306–316, 1985.
37. Bonnet, M.: Shape Differentiation of Regularized BIE: Application to 3D Crack Analysis by the Virtual Crack Extension Approach. In *Boundary Elements in Mechanical and Electrical Engineering.* (Brebbia, C.A.; Chaudouet, A., Eds.), Springer–Verlag, 1990.

38. Bonnet, M.: On the Use of Geometrical Symmetry in the Boundary Element Method for 3D Elasticity. In *Boundary Element Technology VI.* (Brebbia, C.A., Ed.), Southampton, Computational Mechanics Publication, 185–201, 1991.
39. Bowen, R.M.: Theory of Mixtures. In *Continuum Physics.* (Eringen, A.C., Ed.), Vol. III, Academic Press, New York, 1–127, 1976.
40. Brebbia, C.A.; Domínguez, J.: Boundary Element Methods for Potential Problems. *Applied Mathematical Modeling*, **1**, 372–378, 1977.
41. Brebbia, C.A.; Telles, J.C.F.; Wrobel, L.C.: *Boundary Element Techniques.* Springer-Verlag, Berlin, New York, 1984.
42. Bronstein, I. N.; Semendjajew, K. A.: *Taschenbuch der Mathematik.* Harri Deutsch Verlag, Thun, Frankfurt/Main, 21. edition, 1984.
43. Caputo, M.: Vibrations of an Infinite Plate with a Frequency Independent Q. *Journal of the Acoustical Society of America*, **60**(3), 634–639, 1976.
44. Carini, A.; Diligenti, M.; Maier, G.: Boundary Integral Equation Analysis in Linear Viscoelasticity: Variational and Saddle Point Formulations. *Computational Mechanics*, **8**, 87–98, 1991.
45. Chen, J.: Time Domain Fundamental Solution to Biot's Complete Equations of Dynamic Poroelasticity. Part I: Two-Dimensional Solution. *International Journal of Solids and Structures*, **31**(10), 1447–1490, 1994.
46. Chen, J.: Time Domain Fundamental Solution to Biot's Complete Equations of Dynamic Poroelasticity. Part II: Three-Dimensional Solution. *International Journal of Solids and Structures*, **31**(2), 169–202, 1994.
47. Chen, J.; Dargush, G.F.: Boundary Element Method for Dynamic Poroelastic and Thermoelastic Analysis. *International Journal of Solids and Structures*, **32**(15), 2257–2278, 1995.
48. Cheng, A. H.-D.; Badmus, T.; Beskos, D.E.: Integral Equations for Dynamic Poroelasticity in Frequency Domain with BEM Solution. *Journal of Engineering Mechanics ASCE*, **117**(5), 1136–1157, 1991.
49. Cheng, A. H.-D.; Detournay, E.: On Singular Integral Equations and Fundamental Solutions of Poroelasticity. *International Journal of Solids and Structures*, **35**(34-35), 4521–4555, 1998.
50. Cheng, A. H.-D.; Ligget, J.A.: Boundary Integral Equation Method for Linear Porous-elasticity with Applications to Fracture Propagation. *International Journal for Numerical Methods in Engineering*, **20**(2), 279–296, 1984.
51. Cheng, A. H.-D.; Ligget, J.A.: Boundary Integral Equation Method for Linear Porous-elasticity with Applications to Soil Consilidation. *International Journal for Numerical Methods in Engineering*, **20**(2), 255–278, 1984.
52. Cheng, A. H.-D.; Predeleanu, M.: Transient Boundary Element Formulation for Linear Poroelasticity. *Journal of Applied Mathematical Modelling*, **11**, 285–290, 1987.
53. Cheng, A. H.-D.; Sidauruk, P.; Abousleiman, Y.: Approximate Inversion of the Laplace Transform. *The Mathematica Journal*, **4**(2), 76–82, 1994.
54. Cheng, A.H.-D.; Antes, H.: On Free Space Green's Function for High Order Helmholtz Equations. In *Boundary Element Methods: Fundamentals and Applications.* (Kobayashi, S.; Nishimura, N., Eds.), Berlin, Springer-Verlag, 67–71, 1991.
55. Christensen, R.M.: *Theory of Viscoelasticity.* Academic Press, New York, 1971.
56. Coda, H.B.; Venturini, W.S.: Three-Dimensional Transient BEM Analysis. *Computers & Structures*, **56**(5), 751–768, 1995.
57. Creus, G.J.: *Viscoelasticity – Basic Theory and Applications to Concrete Structures*, Vol. 16, *Lecture Notes in Engineering*. Springer, Berlin, Heidelberg, 1986.
58. Crump, K. S.: Numerical Inversion of Laplace Transforms Using a Fourier Series Approximation. *Journal of the Association for Computing Machinery*, **23**(1), 89–96, 1976.

59. Cruse, T.A.: A Direct Formulation and Numerical Solution of the General Transient Elastodynamic Problem, II. *Journal of Mathematical Analysis and Applications*, **22**, 341–355, 1968.
60. Cruse, T.A.: Numerical Solutions in Three-Dimensional Elastostatics. *International Journal of Solids and Structures*, **5**, 1259–1274, 1969.
61. Cruse, T.A.; Rizzo, F.J.: A Direct Formulation and Numerical Solution of the General Transient Elastodynamic Problem, I. *Journal of Mathematical Analysis and Applications*, **22**, 244–259, 1968.
62. de Boer, R.: *Theory of Porous Media*. Springer-Verlag, Berlin, 2000.
63. de Boer, R.; Ehlers, W.: A Historical Review of the Formulation of Porous Media Theories. *Acta Mechanica*, **74**, 1–8, 1988.
64. de Boer, R.; Ehlers, W.: The Development of the Concept of Effective Stresses. *Acta Mechanica*, **83**, 77–92, 1990.
65. de Boer, R.; Ehlers, W.; Liu, Z.: One-Dimensional Transient Wave Propagation in Fluid-Saturated Incompressible Porous Media. *Archive of Applied Mechanics*, **63**, 59–72, 1993.
66. de Hoop, A.T.: An Elastodynamic Reciprocity Theorem for Linear, Viscoelastic Media. *Applied Scientific Research*, **16**, 39–45, 1966.
67. de Langre, E.; Axisa, F.; Guilbaud, D.: Forced Flexural Vibrations of Beams Using a Time-Stepping Boundary Element Method. In *Boundary Elements in Mechanical and Electrical Engineering*. (Brebbia, C.A.; Chaudouet-Miranda, A., Eds.), Computational Mechanics Publications, Southampton, 139–150, 1990.
68. Debnath, L.: *Integral Transform and their Applications*. CRC Press, 1995.
69. Detournay, E.; Cheng, A. H.-D.: *Fundamentals of Poroelasticity*, Vol. II, *Comprehensive Rock Engineering: Principles, Practice & Projects*, Chapter 5, 113–171. Pergamon Press, 1993.
70. Diebels, S.; Ehlers, W.: Dynamic Analysis of a Fully Saturated Porous Medium Accounting for Geometrical and Material Non-Linearities. *International Journal for Numerical Methods in Engineering*, **39**, 81–97, 1996.
71. Doetsch, G.: *Anleitung zum praktischen Gebrauch der Laplace-Transformation und der Z-Transformation*. R. Oldenbourg Verlag, München, Wien, 1967.
72. Domínguez, J.: Dynamic Stiffness of Rectangular Foundations. Report no. R78-20, Department of Civil Engineering, MIT, Cambridge MA, 1978.
73. Domínguez, J.: Response of Embedded Foundations to Traveling Waves. Report no. R78-24, Department of Civil Engineering, MIT, Cambridge MA, 1978.
74. Domínguez, J.: An Integral Formulation for Dynamic Poroelasticity. *Journal of Applied Mechanics, ASME*, **58**, 588–591, 1991.
75. Domínguez, J.: Boundary Element Approach for Dynamic Poroelastic Problems. *International Journal for Numerical Methods in Engineering*, **35**(2), 307–324, 1992.
76. Domínguez, J.: *Boundary Elements in Dynamics*. Computational Mechanics Publication, Southampton, 1993.
77. Drozdov, A.D.: Fractional Differential Models in Finite Viscoelasticity. *Acta Mechanica*, **124**, 155–180, 1997.
78. Dubner, H.; Abate, J.: Numerical Inversion of Laplace Transforms by Relating them to the Finite Fourier Cosine Transform. *Journal of the Association for Computing Machinery*, **15**(1), 115–123, 1968.
79. Durbin, F.: Numerical Inversion of Laplace Transforms: an Efficient Improvement to Dubner and Abate's Method. *The Computer Journal*, **17**(4), 371–376, 1974.
80. Ehlers, W.: Poröse Medien – ein kontinuumsmechanisches Modell auf der Basis der Mischungstheorie. Forschungsbericht aus dem Fachbereich Bauwesen 47, Universität - GH Essen, 1989.
81. Ehlers, W.: Compressible, Incompressible and Hybrid Two-phase Models in Porous Media Theories. *ASME: AMD-Vol.*, **158**, 25–38, 1993.

82. Ehlers, W.: Constitutive Equations for Granular Materials in Geomechanical Context. In *Continuum Mechanics in Environmental Sciences and Geophysics*. (Hutter, K., Ed.), CISM Courses and Lecture Notes, No. 337, Springer-Verlag, Wien, 313–402, 1993.
83. Ehlers, W.; Kubik, J.: On Finite Dynamic Equations for Fluid-Saturated Porous Media. *Acta Mechanica*, **105**, 101–117, 1994.
84. Eringen, A. C.; Suhubi, E. S.: *Elastodynamics*, Vol. II. Academic Press, New York, San Francisco, London, 1975.
85. Fillunger, P.: Der Auftrieb von Talsperren, Teil I-III. *Österr. Wochenschrift für den öffentlichen Baudienst*, 532–570, 1913.
86. Flügge, W.: *Viscoelasticity*. Springer–Verlag, New York, Heidelberg, Berlin, 1975.
87. Fredholm, I.: Sur une Classe d'Equations Fonctionelles. *Acta Mathematica, Sweden*, **27**, 365–390, 1903.
88. Gaul, L.; Klein, P.; Kempfle, S.: Damping Description Involving Fractional Operators. *Mechanical Systems and Signal Processing*, **5**(2), 81–88, 1991.
89. Gaul, L.; Klein, P.; Plenge, M.: Dynamic Boundary Element Analysis of Foundation Slabs on Layered Soil. In *Boundary Elements X*. (Brebbia, C.A., Ed.), Vol. 4, Southampton, Boston, Computational Mechanics Publications, 29–44, 1988.
90. Gaul, L.; Schanz, M.: BEM Formulation in Time Domain for Viscoelastic Media Based on Analytical Time Integration. In *Boundary Elements XIV*. (Brebbia, C.A.; Dominguez, J.; Paris, F., Eds.), Vol. II, Southampton, Computational Mechanics Publications, 223–234, 1992.
91. Gaul, L.; Schanz, M.: Dynamics of Viscoelastic Solids Treated by Boundary Element Approaches in Time Domain. *European Journal of Mechanics A/Solids*, **13**(4–suppl.), 43–59, 1994.
92. Gaul, L.; Schanz, M.: A Viscoelastic Boundary Element Formulation in Time Domain. *Archives of Mechanics*, **46**(4), 583–594, 1994.
93. Gaul, L.; Schanz, M.: Boundary Element Calculation of Transient Response of Viscoelastic Solids Based on Inverse Transformation. *Meccanica*, **32**(3), 171–178, 1997.
94. Gaul, L.; Schanz, M.: A Comparative Study of Three Boundary Element Approaches to Calculate the Transient Response of Viscoelastic Solids with Unbounded Domains. *Computer Methods in Applied Mechanics and Engineering*, **179**(1-2), 111–123, 1999.
95. Geers, T.L.: *IUTAM Symposium on Computational Methods for Unbounded Domains*. Kluwer Academic Publishers, Dodrecht, Boston, London, 1997.
96. Gel'fand, I. M.; Shilov, G. E.: *Generalized Functions*, Vol. I. Academic Press, New York and London, 1964.
97. Giljohann, D.; Bittner, M.: The Three-Dimensional DtN Finite Element Method for Radiation Problems of the Helmholtz Equation. *Journal of Sound and Vibration*, **212**(3), 383–394, 1998.
98. Graff, K. F.: *Wave Motion in Elastic Solids*. Oxford University Press, 1975.
99. Graffi, D.: Über den Reziprozitätssatz in der Dynamik elastischer Körper. *Ingenieur Archiv*, **22**, 45–46, 1954.
100. Grag, S.K.; Nafeh, A.H.; Good, A.J.: Compressional Waves in Fluid-Saturated Elastic Porous Media. *Journal of Applied Physics*, **45**, 1968–1974, 1974.
101. Guiggiani, M.; Gigante, A.: A General Algorithm for Multidimensional Cauchy Principal Value Integrals in the Boundary Element Method. *Journal of Applied Mechanics, ASME*, **57**, 906–915, 1990.
102. Gurtin, M.E.; Sternberg, E.: On the Linear Theory of Viscoelasticity. *Archive for Rational Mechanics and Analysis*, **11**, 291–356, 1962.
103. Hairer, E.; Nørsett, S.P.; Wanner, G.: *Solving Ordinary Differential Equations I: Nonstiff Problems*. Springer Series in Computational Mathematics. Springer-Verlag, Berlin Heidelberg, 1987.

104. Hairer, E.; Wanner, G.: *Solving Ordinary Differential Equations II:Stiff and Differential-Algebraic Problems.* Springer Series in Computational Mathematics. Springer-Verlag, Berlin Heidelberg, 1991.
105. Hörmander, L.: *Linear Partial Differential Operators.* Springer-Verlag, 1963.
106. Ihlenburg, F.: *Finite Element Analysis of Acoustic Scattering.* Applied Mathematical Sciences. Springer-Verlag, New York Berlin Heidelberg, 1998.
107. Jawson, M.A.: Integral Equation Methods in Potential Theory: I. *Proceedings of the Royal Society of London*, **275**(A), 23–32, 1963.
108. Jonsson, J.: Comments to "Ground Vibration Generated by a Load Moving Along a Railway Track". *Journal of Sound and Vibration*, **236**(2), 359–366, 2000.
109. Karabalis, D.L.; Rizos, D.C.: An Advanced Direct Time Domain Boundary Element Method for 3-D Elastodynamics. In *Boundary Elements XV.* (Brebbia, C.A.; Renics, J.J., Eds.), Vol. 2, 347–361, 1993.
110. Karabalis, D.L.; Rizos, D.C.: Dynamic Analysis of 3-D Foundations. In *Boundary Element Techniques in Geomechanics.* (Manolis, G.D.; Davies, T.G, Eds.), London, Elsevier, 1993.
111. Kausel, E.: Thin-Layer Method: Formulation in the Time Domain. *International Journal for Numerical Methods in Engineering*, **37**(6), 927–941, 1994.
112. Kellogg, O.D.: *Foundations of Potential Theory.* Springer-Verlag, Berlin, 1929.
113. Khutoryanski, N.M.; Sosa, H.: Dynamic Representation Formulas and Fundamental Solutions for Piezoelectricity. *International Journal of Solids and Structures*, **32**, 3307–3325, 1995.
114. Kim, Y.K.; Kingsbury, H.B.: Dynamic Characterization of Poroelastic Materials. *Experimental Mechanics*, **19**, 252–258, 1979.
115. Kobayashi, S.; Kawakami, T.: Application of BE-FE Combined Method to Analysis of Dynamic Interactions Between Structures and Viscoelastic Soil. In *Boundary Elements VII.* (Brebbia, C.A.; Maier, G., Eds.), Vol. I, Berlin, Heidelberg, Springer, 6–3 – 6–12, 1985.
116. Kupradze, V.D.: *Potential Methods in the Theory of Elasticity.* Israel Program for Scientific Translations, Jerusalem, 1965.
117. Lakes, R.S.: *Viscoelastic Solids.* CRC Press, Boca Raton, London, 1999.
118. Lubich, C.: Convolution Quadrature and Discretized Operational Calculus. I. *Numerische Mathematik*, **52**, 129–145, 1988.
119. Lubich, C.: Convolution Quadrature and Discretized Operational Calculus. II. *Numerische Mathematik*, **52**, 413–425, 1988.
120. Lubich, Ch.: On the Multistep Time Discretization of Linear Initial-Boundary Value Problems and their Boundary Integral Equations. *Numerische Mathematik*, **67**, 365–389, 1994.
121. Lubich, Ch.; Schneider, R.: Time Discretization of Parabolic Boundary Integral Equations. *Numerische Mathematik*, **63**, 455–481, 1992.
122. Manolis, G.D.: A Comparative Study on Three Boundary Element Method Approaches to Problems in Elastodynamics. *International Journal for Numerical Methods in Engineering*, **19**, 73–91, 1983.
123. Manolis, G.D.; Beskos, D.E.: Dynamic Stress Concentration Studies by Boundary Integrals and Laplace Transform. *International Journal for Numerical Methods in Engineering*, **17**, 573–599, 1981.
124. Manolis, G.D.; Beskos, D.E.: Integral Formulation and Fundamental Solutions of Dynamic Poroelasticity and Thermoelasticity. *Acta Mechanica*, **76**, 89–104, 1989. Errata [125].
125. Manolis, G.D.; Beskos, D.E.: Corrections and Additions to the Paper "Integral Formulation and Fundamental Solutions of Dynamic Poroelasticity and Thermoelasticity". *Acta Mechanica*, **83**, 223–226, 1990.

126. Mansur, W. J.: *A Time-Stepping Technique to Solve Wave Propagation Problems Using the Boundary Element Method.* Phd thesis, University of Southampton, 1983.
127. Mansur, W.J.; Brebbia, C.A.: Transient Elastodynamics Using a Time-Stepping Technique. In *Boundary Elements.* (Brebbia, C.A.; Futagami, T.; Tanaka, M., Eds.), Berlin, Springer-Verlag, 677–698, 1983.
128. Mansur, W.J.; Carrer, J.A.M.; Siqueira, E.F.N.: Time Discontinuous Linear Traction Approximation in Time-Domain BEM Scalar Wave Propagation. *International Journal for Numerical Methods in Engineering*, **42**(4), 667–683, 1998.
129. Mantič, V.: A new Formula for the C-matrix in the Somigliana Identity. *Journal of Elasticity*, **33**, 191–201, 1993.
130. Meyer, O.E.: Zur Theorie der inneren Reibung. *Journal für die reine und angewandte Mathematik*, **78**, 130–135, 1874.
131. Muskhelishvili, N.I.: *Singular Integral Equations.* Noordhoff Publishing Co., Groningen, 1953.
132. Muskhelishvili, N.I.: *Some Basic Problems of the Mathematical Theory of Elasticity.* Noordhoff Publishing Co., Groningen, 1953.
133. Narayanan, G.V.; Beskos, D.E.: Numerical Pperational Methods for Time-Dependent Linear Problems. *International Journal for Numerical Methods in Engineering*, **18**, 1829–1854, 1982.
134. Nardini, D.; Brebbia, C.A.: A New Approach to Free Vibration Analysis Using Boundary Elements. In *Boundary Element Methods.* (Brebbia, C.A., Ed.), Springer-Verlag, Berlin, 312–326, 1982.
135. Norris, A.N.: Dynamic Green's Functions in Anisotropic Piezoelectric, Thermoelastic and Poroelastic Solids. *Proceedings of the Royal Society of London*, **447**(A), 175–188, 1994.
136. Nowacki, W.: *Dynamic Problems of Thermoelasticity.* Nordhoff, Leyden, 1975.
137. Oldham, K. B.; Spanier, J.: *The Fractional Calculus*, Vol. 111, *Mathematics in Science and Engineering.* Academic Press, New York, London, 1974.
138. Ortner, N.: Die Fundamentallösung des Timoshenko- und des Boussinesq-Operators. *Zeitschrift für angewandte Mathematik und Mechanik*, **68**(11), 547–553, 1988.
139. Pak, R.Y.S.: Asymmetric Wave Propagation in an Elastic Half-Space by a Method of Potentials. *Journal of Applied Mechanics, ASME*, **54**, 121–126, 1987.
140. París, F.; Cañas, J.: *Boundary Element Method: Fundamentals and Applications.* Oxford University Press, Oxford New York Tokyo, 1997.
141. Partridge, P.W.; Brebbia, C.A.; Wrobel, L.C.: *The Dual Reciprocity Boundary Element Method.* Computational Mechanics Publication, Southampton, 1992.
142. Peirce, A.; Siebrits, E.: Stability Analysis and Design of Time-Stepping Schemes for General Elastodynamic Boundary Element Models. *International Journal for Numerical Methods in Engineering*, **40**(2), 319–342, 1997.
143. Pekeris, C.L.: The Seismic Surface Pulse. *Proc. of the National American Society*, **41**, 469–480, 1955.
144. Plona, T.J.: Observation of a Second Bulk Compressional Wave in Porous Medium at Ultrasonic Frequencies. *Applied Physics Letters*, **36**(4), 259–261, 1980.
145. Podlubny, I.: *Fractional Differential Equations*, Vol. 198, *Mathematics in Science and Engineering.* Academic Press, San Diego, New York, London, 1999.
146. Providakis, C.P.; Beskos, D.E.: Dynamic Analysis of Beams by the Boundary Element Method. *Computers & Structures*, **22**(6), 957–964, 1986.
147. Rayleigh, J.W.S.: On Waves Propagated Along the Plane Surface of an Elastic Solid. *Proceedings of the London mathematical Society*, **17**, 4–11, 1887.
148. Rellich, F.: Über das asymptotische Verhalten der Lösungen von $\Delta u + ku = 0$ in unendlichen Gebieten. *Jahresbericht der deutschen Mathematiker Vereinigung*, **53**, 57–64, 1943.

149. Richards, J.I.; Youn, H.K.: *Theory of Distributions: A Non-technical Introduction.* Cambridge University Press, 1990.
150. Rizos, D.C.; Karabalis, D.L.: An Advanced Direct Time Domain BEM Formulation for General 3-D Elastodynamic Problems. *Computational Mechanics*, **15**, 249–269, 1994.
151. Rizzo, F.J.: An Integral Equation Approach to Boundary Value Problems of Classical Elastostatics. *Quarterly of Applied Mathematics*, **25**, 83–95, 1967.
152. Rossikhin, Y.A.; Shitikova, M.V.: Applications of Fractional Calculus to Dynamic Problems of Linear and Nonlinear Hereditary Mechanics of Solids. *Applied Mechanics Review*, **50**(1), 15–67, 1997.
153. Ryshik, I.M.; Gradstein, I.S.: *Summen-, Produkte- und Integraltafeln.* VEB Deutscher Verlag der Wissenschaften, Berlin, 1963.
154. Sáez, A.; Domínguez, J.: BEM Analysis of Wave Scattering in Transversely Isotropic So lids. *International Journal for Numerical Methods in Engineering*, **44**, 1283–1300, 1999.
155. Schanz, M.: A Time Domain Boundary Element Formulation Based on a Multistep Time Discretization. In *Boundary Elements XIX.* (Marchetti, M.; Brebbia, C.A.; Aliabadi, M.H., Eds.), Southampton, Computational Mechanics Publications, 767–776, 1997.
156. Schanz, M.: Multistep Time Discretization of Boundary Integral Equations in Dynamics. In *Modelling and Simulation Based Engineering.* (Atluri, S.N.; O'Donoghue, P.E., Eds.), Vol. I, Palmdale, Tech Science Press, 223–228, 1998.
157. Schanz, M.: A Boundary Element Formulation in Time Domain for Viscoelastic Solids. *Communications in Numerical Methods in Engineering*, **15**, 799–809, 1999.
158. Schanz, M.: Boundary Element Calculation of Wave Propagation in 3-d Poroelastic Solids. In *Advances in Computational Engineering & Sciences.* (Atluri, S.N.; Brust, F.W., Eds.), Vol. I, Palmdale, Tech Science Press, 118–123, 2000.
159. Schanz, M.: Application of 3-d Boundary Element Formulation to Wave Propagation in Poroelastic Solids. *Engineering Analysis with Boundary Elements*, (in press).
160. Schanz, M.: Dynamic Poroelasticity Treated by a Time Domain Boundary Element Method. *IUTAM Symposium on Advanced Mathematical and Computational Mechanic Aspects of the Boundary Element Method.* Kluwer Academic Publishers, (in press).
161. Schanz, M.; Antes, H.: Application of 'Operational Quadrature Methods' in Time Domain Boundary Element Methods. *Meccanica*, **32**(3), 179–186, 1997.
162. Schanz, M.; Antes, H.: A New Visco- and Elastodynamic Time Domain Boundary Element Formulation. *Computational Mechanics*, **20**(5), 452–459, 1997.
163. Schanz, M.; Cheng, A.H.-D.: Dynamic Analysis of a One-Dimensional Poroviscoelastic Column. *Journal of Applied Mechanics, ASME*, (in press), 2000.
164. Schanz, M.; Cheng, A.H.-D.: Transient Wave Propagation in a One-Dimensional Poroelastic Column. *Acta Mechanica*, **145**, 1–18, 2000.
165. Schanz, M.; Gaul, L.: Implementation of Viscoelastic Behaviour in a Time Domain Boundary Element Formulation. *Applied Mechanics Review*, **46**(11, Part 2), S41–S46, 1993.
166. Schanz, M.; Gaul, L.; Antes, H.: Numerical Damping and Instability of a 3–D BEM Time–Stepping Algorithm. *Extended Abstracts of IABEM 93*, Braunschweig, 1993.
167. Schanz, M.; Gaul, L.; Wenzel, W.; Zastrau, B.: A Boundary Element Formulation for Generalized Viscoelastic Solids in Time Domain. In *Boundary Element Topics.* (Wendland, W., Ed.), Springer, Heidelberg, 31–50, 1997.
168. Schwartz, L.: *Théorie des distributions.* Herman, Paris, 1966.
169. Sheng, X.; Jones, C.J.C.; Petyt, M.: Ground Vibration Generated by a Load Moving Along a Railway Track. *Journal of Sound and Vibration*, **228**(1), 129–156, 1999.
170. Sim, W.J.; Kwak, B.M.: Linear Viscoelastic Analysis in Time Domain by Boundary Element Method. *Computers & Structures*, **29**(4), 531–539, 1988.

171. Sommerfeld, A.: *Partial Differential Equations in Physics*. Academic Press, New York, 1949.
172. Spiegel, M. R.: *Laplace Transforms*. Schaum's Outline. McGraw–Hill Book Company, Düsseldorf, New–York, 1965.
173. Stakgold, I.: *Green's Functions and Boundary Value Problems*. Pure and Applied Mathematics. John Wiley & Sons, 2nd edition, 1998.
174. Suiker, A.S.J; de Borst, R.; Esveld, C.: Critical Behaviour of a Timoshenko Beam-Half Plane System Under a Moving Load. *Archive of Applied Mechanics*, **68**, 158–168, 1998.
175. Tanaka, M.: *New Integral Equation Approach to Viscoelastic Problems*, Vol. 3: Computational Aspects, *Topics in Boundary Element Research*, Chapter 2, 25–35. Springer, Berlin, Heidelberg, 1987.
176. Torvik, P.J.; Bagley, R.L.: Fractional Derivatives in the Description of Damping Materials and Phenomena. In *The Role of Damping in Vibration and Noise Control.* (Rogers, L.; Simonis, J.C., Eds.), The American Society of Mechanical Engineers, 125–135, 1987. DE–Vol. 5.
177. Triantafyllidis, Th.: 3-D Time Domain BEM Using a Half-Space Green's Functions. *Engineering Analysis with Boundary Elements*, **8**, 115–124, 1991.
178. Truesdell, C.; Toupin, R.A.: The Classical Field Theories. In *Handbuch der Physik.* (Flügge, S., Ed.), Vol. III/1, Springer-Verlag, Berlin, 226–793, 1960.
179. Vardoulakis, I.; Beskos, D.E.: Dynamic Behavior of Nearly Saturated Porous Media. *Mechanics of Composite Materials*, **5**, 87–108, 1986.
180. Vgenopoulou, I.; Beskos, D.E.: Dynamic Behavior of Saturated Poroviscoelastic Media. *Acta Mechanica*, **95**, 185–195, 1992.
181. von Terzaghi, K.: Die Berechnung der Durchlässigkeit des Tones aus dem Verlauf der hydromechanischen Spannungserscheinungen. *Sitzungsbericht der Akademie der Wissenschaften (Wien): Mathematisch–Naturwissenschaftlichen Klasse*, **132**, 125–138, 1923.
182. Wang, C.Y.; Achenbach, J.D.: Three-Dimensional Time-Harmonic Elastodynamic Green's Functions for Anisotropic Solids. *Proceedings of the Royal Society of London*, **449**(A), 441–458, 1995.
183. Wang, C.Y.; Achenbach, J.D.; Hirose, S.: Two-Dimensional Time Domain BEM for Scattering of Elastic Waves in Solids of General Anisotropy. *International Journal of Solids and Structures*, **33**(26), 3843–3864, 1996.
184. Wiebe, Th.; Antes, H.: A Time Domain Integral Formulation of Dynamic Poroelasticity. *Acta Mechanica*, **90**, 125–137, 1991.
185. Wolf, J.P.; Dabre, G.R.: Time-Domain Boundary Element Method in Viscoelasticity with Application to a Spherical Cavity. *Soil Dynamics and Earthquake Engineering*, **5**, 138–148, 1986.
186. Yu, G.; Mansur, W.J.; Carrer, J.A.M.: A Linear θ for 2-D Elastodynamic BE Analysis. *Computational Mechanics*, **24**, 82–89, 1999.
187. Yu, G.; Mansur, W.J.; Carrer, J.A.M.; Gong, L.: A Linear θ Method Applied to 2D Time Domain BEM. *Communications in Numerical Methods in Engineering*, **14**(12), 1171–1179, 1998.
188. Yu, G.; Mansur, W.J.; Carrer, J.A.M.; Gong, L.: Time Weighting in Time Domain BEM. *Engineering Analysis with Boundary Elements*, **22**(3), 175–181, 1998.
189. Yu, G.; Mansur, W.J.; Carrer, J.A.M.; Gong, L.: Stability of Galerkin and Collocation Time Domain Boundary Element Methods as Applied to the Scalar Wave Equation. *Computers & Structures*, **74**(4), 495–506, 2000.
190. Zienkiewicz, O.C.: *The Finite Element Method.* McGraw-Hill Book Company, 1977.
191. Zienkiewicz, O.C.; Paul, D.K.; Chan, A.H.C.: Unconditionally Stable Staggered Solution Procedure for Soil-Pore Fluid Interaction Problems. *International Journal for Numerical Methods in Engineering*, **26**, 1039–1055, 1988.

192. Zienkiewicz, O.C.; Shiomi, T.: Dynamic Behaviour of Saturated Porous Media; The Generalized Biot Formulation and its Numerical Solution. *International Journal for Numerical and Analytical Methods in Geomechanics*, **8**, 71–96, 1984.

Index

Printing (Computer to Film): Saladruck, Berlin
Binding: Stürtz AG, Würzburg